Templets and the explanation
of complex patterns

Templets and the explanation of complex patterns

MICHAEL J. KATZ

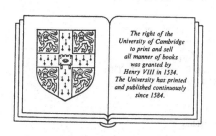

The right of the
University of Cambridge
to print and sell
all manner of books
was granted by
Henry VIII in 1534.
The University has printed
and published continuously
since 1584.

CAMBRIDGE UNIVERSITY PRESS

CAMBRIDGE

LONDON   NEW YORK   NEW ROCHELLE

MELBOURNE   SYDNEY

CAMBRIDGE UNIVERSITY PRESS
Cambridge, New York, Melbourne, Madrid, Cape Town, Singapore, São Paulo, Delhi

Cambridge University Press
The Edinburgh Building, Cambridge CB2 8RU, UK

Published in the United States of America by Cambridge University Press, New York

www.cambridge.org
Information on this title: www.cambridge.org/9780521306737

First published 1986
This digitally printed version 2008

A catalogue record for this publication is available from the British Library

Library of Congress Cataloguing in Publication data

Katz, Michael J.
Templets and the explanation of complex patterns.
Includes index.
1. Biochemical templetes. 2. Morphology—
Philosophy. 3. Biological control systems. I. Title.
QP517.B48K38   1985   574.19′2   85–11304

ISBN 978-0-521-30673-7 hardback
ISBN 978-0-521-09602-7 paperback

Dedication

For my parents
Sidney Katz and Beverly Suid Katz

Contents

Prologue

In this work, I owe a special debt to two colleagues. Raymond J. Lasek, from whom I learned biology, always provided a thoughtful, challenging, and critical response as I presented him with various steps in my thinking out of these ideas. Ulf Grenander was my warm and generous host in the Division of Applied Mathematics at Brown University. His Pattern Theory is the language that I have borrowed in an attempt to formalize my ideas about templeting. A language is more than a tool, it structures one's thoughts; and Prof. Grenander's language is responsible for much of the organization of my ideas. Both of these scientists engaged me in innumerable hours of epistemological discussion, and I cannot help but to have adopted many of their good ideas. Some of these are now woven inextricably throughout this book.

I would also like to thank the Alfred P. Sloan Foundation and the Whitehall Foundation, which provided financial support for this work.

The title of Chapter 2 is borrowed whole from the book *The Nature Of Explanation* in which the author, Kenneth Craik, makes a strong case for the importance of determinate explanations.

The story that I present in this book is not as tidy as I had originally hoped, but I think that this is the nature of detailed abstractions that must be tethered to the real world. Hermann Weyl introduces his *Philosophy Of Mathematics And Natural Science* as follows:

With the years, I have grown more hesitant about the metaphysical implications of science; 'as we grow older, the world becomes stranger, the pattern more complicated'.

(Here, Weyl is quoting from T.S. Eliot's 'Four Quartets: East Coker, V':

Home is where one starts from. As we grow older
The world becomes stranger, the pattern more complicated
Of dead and living.)

In other words, the crystalline metaphysics that jumps out at one in the boldly colored, pristine vistas of youth becomes unsettlingly complex over time. That is my reading of Weyl's statement, and it has also been my experience.

M.J. Katz, 1985

Introduction

Where am I? Metaphysics says
No question can be asked unless
It has an answer, so I can
Assume this maze has got a plan.

(W.H. Auden 'The labyrinth'.)

How can a scientist make sense of very complex systems? What logical abstractions can·he use to dissect phenomena that cannot easily be reduced to simpler parts? Like all scientists, researchers faced with complex systems wish to construct rigorous scientific explanations; yet the particular phenomena that they have at hand are often forbiddingly tangled. Some complex phenomena are so heterogeneous that they appear almost random; but, surprisingly, they differ from the truly random in that they are faithfully reproducible. 'Random' means more than extremely heterogeneous – it also means unpredictable.

In most methods for analyzing random phenomena, these two ideas – heterogeneity and unpredictability – are inextricably intertwined. For this reason, the standard stochastic tools for explaining random phenomena often cannot be effectively applied to heterogeneous yet highly stereotyped (predictable) phenomena. The realm of the very heterogeneous is not the usual realm of classical physics. But, the task of scientifically explaining heterogeneous phenomena is the task of most other disciplines. It represents the daily challenges of biology, psychology, sociology, economics, meteorology, epidemiology, metallurgy and geology. Theoreticians in each of these disciplines continually face inescapable, and at times irreducible, heterogeneity. And, irreducible heterogeneity is tantamount to complexity.

This book presents the beginnings of one epistemological approach to the natural science of complex systems. In the following pages, I argue that one of the jobs of any science is to build a certain type of abstract model of configurations of real world items. These configurations are 'patterns'. It is beyond the scope of the present work to face the wonderful problem of how a scientist decides what real world entities can be considered respectable items and which real world configurations can be demarcated as interesting patterns. Such decisions are probably a mixture of intuition, insight, historical accident, cultural structure, and even the intrinsic nature of the human nervous system. Suffice it to say that whatever the basis for a scientist's decision, he begins by identifying certain real world patterns.

The scientific task, then, is to explain these patterns. A scientific explanation is a toy model with a particular structure, and it can be captured on paper as a precisely interlinked collection of abstractions. Specifically, the real world pattern that is to be explained is represented by one abstraction, and this abstraction is carefully tied – through a determinate transforming operation – to a set of predecessor abstractions. As Carnap has pointed out, the predecessor abstractions should include both a 'universal law' and a class of particular situational constraints. In turn, the broader constructure of science is built of such scientific explanations assembled into larger explanations, scientific theories.

Most of this book is devoted to exploring one class of scientific explanations, the configurational explanations. Configurational explanations are concerned with the organization (the topology) of a pattern rather than with its content. These explanations may prove especially useful in making sense of the forms of those complex systems, such as the biological systems, that have been refractory to the usual theoretical analyses.

Throughout the text, I have tried to use biological examples. Living systems are the archetypic complex systems in the natural world: in fact, as Schrodinger has written, the faithful recreation of complex patterns can be considered to be the hallmark of the biological realm. Ultimately, a theoretical discipline that makes sense of the truly biological phenomena, those phenomena such as ontogeny and phylogeny (development and evolution) that uniquely characterize the biological realm, must base itself in the analysis of the dynamics of complex patterns. Theoretical biology must evolve into a *pattern biology*, a biology concerned with the organization of forms:

Men talk much of matter and energy, of the struggle for existence that molds the shape of life. These things exist, it is true; but more delicate, elusive, quicker than the fins in water, is that mysterious principle known as 'organization,' which leaves all other mysteries concerned with life stale and insignificant by comparison. For that without organization life does not persist is obvious. Yet this organization itself is not strictly the product of life, nor of selection. Like some dark and passing shadow within matter, it cups out the eyes' small windows or spaces the notes of a meadow lark's song in the interior of a mottled egg. That principle – I am beginning to suspect – was there before the living in the deeps of water.

> [L. Eiseley (1956) 'The flow of the river', *The Immense Journey* Random House, NY.]

Configurational explanations make a clean split between the types of topological information used to construct a given pattern. On the one hand, there is the topological information that is inherent in the unordered (or, more accurately, the arbitrarily ordered) set of precursor elements – the raw materials; this is the 'universal law' in a scientific explanation. On the other hand, there is all of the remaining topological information. This, the additional global information, is represented as a blueprint or templet. The heart of a configurational explanation is its templet – the specific situational constraints in a scientific explanation.

In a configurational explanation, a set of precursor elements and a templet are combined by a simple precise operation, logical addition, to yield the final pattern. Configurational explanations are equivalent to basic automata that perform the logical operation *and*. Configurational explanations partition the pattern assembly process into two independent transformations. Within the configurational explanation, topology (organization) is transformed and content is held constant. Between configurational explanations – in scientific theories, which are concatenations of configurational explanations – content is transformed and topology can be held constant.

The elemental philosophical postulate underlying configurational explanations is that order cannot be created *ex nihilo*, a principle that antedates Aristotle. Order must come from preexisting order, and the templet embodies the requisite preexisting order of any pattern. At times, the operant templets may not be easy to discern in the real world, and in stochastic processes the templets are impermanent – they are ephemeral entities that must be recreated at each using. Nonetheless, in the abstract realm of scientific explanation, the explicit identification of a pattern's templet provides a standard means for making sense of complex systems.

It is often assumed that the only 'true' or, at least, the only satisfactory explanations of natural phenomena reduce the phenomena to smaller, simpler or less intricate elements. Configurational explanations belie this hope, clearly demonstrating that this is not always the case. It is not always possible to explain patterns in terms of simpler patterns – one must sometimes invoke complex precursors. Complexity is oft begotten of complexity, and for this reason there must be some 'true', satisfying, and scientifically useful explanations which are based upon complexity and which cannot be entirely reduced to simpler terms. Configurational explanations offer one possible scheme for organizing such complex scientific explanations; and, within configurational explanations, the natural unit for the reduction of the form of a pattern is the templet.

Einstein has written:

The reciprocal relationship of epistemology and science is of noteworthy kind. They are dependent upon each other. Epistemology without contact with science becomes an empty scheme. Science without epistemology is – insofar as it is thinkable at all – primitive and muddled.

> [(1949) 'Remarks concerning the essays brought together in this co-operative volume', *Albert Einstein: Philosopher–Scientist* (P.A. Schlipp, ed.) Open Court, La Salle IL, pp. 683–4.]

Epistemology and science are tied in an iterative reciprocal interaction. This intertwining of the abstract with the real imposes a standard on any epistemology of the natural sciences and makes it possible to argue that some epistemologies are better than others. The essential criterion for 'better' is usefulness, and a scientific epistemology should be judged by whether it allows the scientist to forge new connections between extant ideas, by whether it suggests new and simpler synthetic statements, and by whether it provides a rationale for designing new experiments, perhaps even entirely new varieties of experiment. Most importantly, an epistemology is useful if it extends the horizons of science. When it is first presented, an epistemology should make it possible for a scientist to say: 'I can now see a number of new things, I now have a number of new ideas, and, for the moment, I cannot see the end of this newness.'

1

Scientific abstractions

Abstractions are mental constructs whose main property is that discourse on them is possible without pointing to objects in the external world. In theoretical science the abstractions serve, however, as images of external things.

[W.M. Elsasser (1975) *The Chief Abstractions Of Biology* Elsevier, NY, p. 4.]

Abstractions are for pockets; they are miniatures of the world that we can carry around with us, that we can take out at our leisure and examine, and that we can tinker with. We can poke them and probe them and rearrange their parts. In essence, they are pocket toys.

Abstractions are pocket models of the world, and the scientific abstractions are a special class of these pocket models. For the scientist, abstractions must be *useful* models of the real world – the scientist would like to ensure that what he learns from tinkering with an abstraction will lead him to understand parts of the real world that he has not directly put his hands on. Scientific abstractions must have tiny portals that are windows into the unknown. For this to hold true, for a scientist's abstractions to enable him to see beyond his reference texts, he must be able to generalize from observations and experiments on the abstraction to observations and experiments in the real world. Thus, certain relations must exist between the abstraction and the real world phenomena and certain internal relations must also exist between the parts of the abstraction.

What are these relations? Let me begin the answer to this question by examining a specific example. Consider a stick figure (Fig. 1), the abstraction of a person. The elements of the abstraction are the abstract raw materials that we have used to build our pocket model: the stick hands, the stick arms, the stick legs, the stick trunk, the stick head, the stick eyes, the stick ears, the stick nose and the stick mouth. The corresponding real world items are a person's hands, arms, legs, trunk, head, eyes, ears, nose and mouth. Stick figures

Fig. 1

are quite portable – we can carry them around anywhere, and they are infinitely tinkerable – we can bend and fold them, we can rearrange their parts, and we can concatenate them and dissect them. Most importantly, they can be used to discover details about the workings of the real world, even details that we may not have directly explored.

A one-to-one mapping between elements of a scientific abstraction and items in the real world

What are the key features of the stick figure that make it into a scientific abstraction? First, the elements of the abstraction – the parts of the stick figure body – can be explicitly tied to particular things out in the real world: they can be put in a one-to-one relation with various items of interest in an actual pattern. This one-to-one relation, a 'one-to-one mapping', means that for every stick part in our abstraction there is some unique and specified item of interest in the real world and, conversely, that for every item of interest in the real world some distinct stick part has been specified. One-to-one mappings, as H. Weyl [(1949) *Philosophy Of Mathematics And Natural Science*, Princeton Univ. Press] and J. Piaget [(1971) *Biology and Knowledge*, Univ. Chicago Press] have pointed out, are elemental features of useful scientific abstractions.

The fact that there is a one-to-one mapping between elements of the abstraction and items in the real world means that there should be no ambiguity either when matching the abstraction to the world or when matching the world to the abstraction. As an important corollary, the one-to-one mapping implies that the abstraction must be constructed of discrete elements. This requirement – that the abstraction be discrete – stems from the coincidence of two fundamental characteristics of any one-to-one mapping.

A one-to-one mapping is a binary matching operation. In our case, either an element of the abstraction is definitely the match to a particular item in the real world or it is definitely not. We are not allowed to be wishy-washy, and there are no half-way relations and no variable affiliations. In addition, the one-to-one mapping means that more than one element of the abstraction cannot be matched to a single item in the real world, and *vice versa*. Overlapping matches are forbidden.

Together, the binary nature of the matching operation and the proscription of overlapping matches require us to be able to definitively distinguish one element of our abstraction from the others. We must be able to put each element of the abstraction into a definite and different bin according to its match with the real world. Therefore, we must be able to break up our abstraction into separate pieces, each with a unique identity. In turn, this means that we must put definite boundaries between elements of the abstraction, so that the elements are clearly distinct. An abstraction in which the elements have definite boundaries or edges is a discrete abstraction.

People have a penchant for breaking the world into quanta, discrete packets that we call 'things' or 'items'. When studying the world, we especially attend to edges and boundaries. We see shapes in a heavy mist; we put edges on clouds. We are activated by change and lulled by continua. The epistemological abstraction of nature as composed of simple binary phenomena is a philosophical stance inherited from the far reaches of antiquity. Aristotle, for instance, begins his *Natural Science* by writing:

All of the earlier philosophers agree in taking pairs of opposites as basic principles.

And as Wallace Stevens wrote (in *The glass of water*):

> That the glass would melt in heat,
> That the water would freeze in cold,
> Shows that this object is merely a state,
> One of many, between two poles. So,
> In the metaphysical, there are these poles.

This perceptional digitization of the world is built into our nervous systems; as a rule, nerve cells respond not simply to stimuli but rather to changes in stimuli and they react most strongly to sharp discontinuities in stimuli. In our perception, we do not float amidst interwoven continua, we jostle in a sea of discrete items.

Beyond this the human is also a neat and tidy world-builder. He is not comfortable swimming in a disordered mass of units that have no global organization, that fluidly make and break connections, and that form only the most evanescent complexes. Humans have an inherent need for order, and so we arrange the items about us into particular configurations or 'patterns': 'everything in the world must have design or the human mind rejects it' [J. Steinbeck (1962) *Travels With Charley* Bantam, NY, p. 63]. Thus, as an epistemological foundation, as one fundamental philosophical base for natural science, it is probably fair to say: 'We will consider the real world to be formed of particular configurations of discrete items – i.e., finite discrete patterns.'

In its emphasis on the discrete and finite nature of natural patterns, this simple beginning channels any formal metaphysical analyses along a special set of pathways – those of discrete mathematics. In the natural sciences, these paths have been infrequently followed. Beginning with Newton, complex theoretical contributions to the natural sciences have usually been based on continuum mathematics – such as differential calculus and the geometries and the algebras of continua (essentially, 'surfaces') – the most powerful of mathematical tools. The mathematics of continua are based on the assumption that the substance of the world varies in a smooth and uniform way at the ultramicroscopic level. For this reason, the formal characterization of some property of the substance can be predictably extended over great distances by defining that property at the ultramicroscopic level and then following its changes in infinite tiny automated steps.

Finite discrete patterns do not easily lend themselves to such continuum mathematics. The discrete nature of these patterns means that there will be dramatic discontinuities at the microscopic level, as one encounters the boundaries between elements. Moreover, the finite nature of the patterns means that these edges cannot (in most cases) be smoothed out by stepping back so far – by examining things in the limit – that the discontinuities become imperceptible.

It is, therefore, thanks to the approximate homogeneity of the matter studied by physicists, that mathematical physics came into existence. In the natural sciences the following conditions are no longer to be found: – homogeneity, relative independence of remote parts, simplicity of the elementary fact; and that is why the student of natural science is compelled to have recourse to other modes of generalisation.

[H. Poincare (1952) *Science And Hypothesis*, Dover, NY, p. 159.]

Finite discrete patterns can be most directly formalized by discrete mathematics, the fields of which include: graph theory, pattern theory, automata theory and other discrete branches of computer science, discrete logic, and combinatorics.

Preservation of topology in the mapping between a scientific abstraction and the real world

A one-to-one mapping is the critical relation between the elements of a scientific abstraction and the items of interest in the real world. In addition, useful scientific abstractions have their elements arranged in a particular way within the abstraction. It is not sufficient for us to have an appropriate set of elements in our model if the arrangement of those elements is incompatible with the arrangement of their real world counterparts.

In our stick figure abstraction, the stick hands, the stick legs, the stick head, the stick eyes, and so on, are all arranged naturally, and this permits us to learn about real people by examining our model. For instance, we might ask: What parts of the body must the right hand cross over to touch the left ear? From our abstraction, we can hypothesize that the right hand will cross over the chest, the neck, and the left cheek, and experiments in the real world confirm that this is indeed a good hypothesis.

In contrast, imagine that a tornado has struck our stick figure. Now, even if we still have all of the parts of a real person represented in the model, when those parts are inappropriately or chaotically arranged we will produce poor hypotheses about the real world. After a tornado, the stick hand may well have to cross the stick foot to reach the stick ear.

The stick figure immediately suggests that the relations between the elements of a scientific abstraction must faithfully mirror important associations between their companion elements in the real world – the order in the abstraction must follow the order of the real world. One fundamental aspect of the order of an abstract structure is its topology, and the topology of a scientific abstraction must reflect a comparable topology in the real world. But, what exactly is the topology of a discrete abstraction and how can we construct our abstraction so that it mirrors the topology of the real world?

The topology of any structure is the neighbor relations of that structure – essentially, it is who is connected to whom. Frequently, topology is thought of in the context of surfaces, where a topology-preserving transformation is one that leaves neighboring points still neighbors. When we transform a surface by stretching it or bending it, we nonetheless preserve its topology: these operations can change the area of a surface, but they will not change the adjacency relations between any two points. The famous trick of stretching and bending a coffee cup into the shape of a doughnut is an example of such a topology-preserving transformation.

In a similar sense, the topology of a discrete structure is the adjacency or neighbor relations of its elements, and a topology-preserving transformation of a discrete structure is one that leaves the neighbor relations intact. Gentle bending and folding of a pearl necklace does not change the connections between adjacent pearls.

To have a well-defined topology for a discrete structure, one must have well-defined neighbor relations, and thus, as a beginning, any scientific abstraction must include an explicit description of the neighbor relations or interconnections between its elements. For a useful scientific abstraction, a well-defined topology is the foundation – it is necessary but not sufficient: in addition, the abstract topology must also reproduce certain real world topological relations. Operationally, this means: First, we must identify some important relation among items in the real world. Next, we must build our abstraction so that if two items in the real world share that particular relation then the corresponding elements in our abstraction will be interconnected – i.e., the elements will be topological neighbors. In a useful scientific abstraction, there must be a concordance of topology between the abstract realm and the physical realm – i.e., we must ensure an isomorphism between the abstract topology and the real topology.

The brain as a scientific abstractor

The twentieth century has seen dramatic changes in the epistemology of science. Most notably, in the face of Gödel's famous incompleteness proof, it has become difficult to argue that there is a perfect or a right method for building scientific abstractions. Nonetheless, there are better methods and there are worse methods, and 'better' means 'most useful'. Commonly, useful scientific abstractions have been built on two principles: discrete one-to-one mappings and isomorphism of topology [e.g., A.S. Eddington (1929) *The Nature Of The Physical World* Macmillan, NY; M.R. Cohen & E. Nagel (1934) *An Introduction To Logic And Scientific Method* Harcourt Brace, NY; P.W. Bridgman (1936) *The Nature Of Physical Theory* Princeton Univ. Press, Princeton; H. Weyl (1949) *Philosophy Of Mathematics And Natural History* Princeton Univ. Press, Princeton]. Why do these principles give rise to useful abstractions, explanations and theories? Perhaps it is because our brains work under quite similar principles.

The brain operates directly on representations – abstractions – of the real world, and these abstractions are always spatial maps. The visual world is represented as a spatial map of changes of light intensity, the somatosensory world is represented as a spatial map of the physical stimuli impinging on our bodies, the auditory world is represented as a spatial map of the sound frequencies about us, even the olfactory world appears to be translated into a spatial map.

These maps are physically laid out in specific populations of nerve cells located throughout the nervous system. The adult brain comes with preconceived notions; it has a good idea about what to expect from the world, and much of the spatial detail of its maps is already wired into the neural populations. Rather than continually building full-scale maps anew, external stimuli, relayed to the nervous system from the environment, highlight particular subsets of extant maps. The highlighting – turning on and off particular nerve cells and modulating particular synapses – is conveyed through chemical and electrical impulses travelling to and from each neural population along the interconnecting axon highways. Although many details of the signalling between nerve cells are analogue – i.e., variable along a continuum – other details are clearly digital, and the overall effect can usually be approximated as quantal. Nerve cells tend to be most strongly affected by *changes* in stimuli, and in this way neurons are particularly attuned to edges, to boundaries, and to discrete steps. The result is that the neural maps tend to discretize the world, and the brain abstractions can usually be read in terms of a one-to-one mapping with particular items of the real world. Because of the inherent nature of our brains, we naturally break the world into discrete bits. Although

> Something there is that doesn't love a wall,
> That sends the frozen-ground-swell under it,
> And spills the upper boulders in the sun;
>
> [Robert Frost 'Mending wall']

something there also is within us that continually builds and rebuilds those walls.

Moreover, the information flow between neural populations (between the various brain abstractions) usually preserves topology – as a first approximation, the mappings are topologically isomorphic. In this way, the spatial map of a set of sensory impulses can still be recognized many layers deep in the brain. The nervous system is a hierarchy of topologically matched populations of nerve cells, and an inner neural population decodes the information that it receives largely on the basis of the topological relations of the incoming signals. The universal language of the brain is not a particular chemical, it is spatial maps. With this type of informational organization, central processing is usually the geometric distortion and the differential weighting of topologically conserved abstractions. The innermost brain abstractions are bent, folded, stretched, merged and highlighted, but they tend to remain coherent. Deep in the brain, Pablo Picasso, not Jackson Pollock, is actively at work on our neural maps.

Elegance

Of all the pocket models that a scientist carries around with him, the scientific abstractions form a special class. Scientific abstractions are built of discrete elements, and the interconnections between those elements are explicitly specified. For the abstraction to be useful, there must be a one-to-one mapping between the elements of the abstraction and some set of items in the real world, and there must also be a corresponding one-to-one mapping between the topology – the set of interconnections – of the abstraction and a comparable topology of the real world.

Another less tangible characteristic is sometimes added to these fundamental requirements for any scientific abstraction – namely, elegance. 'Elegance' can mean 'aesthetic appeal', and in this sense science should undoubtedly follow Einstein, who wrote:

I conscientiously adhered to the rule of the gifted theoretician L. Boltzmann – that matters of elegance should be left to the tailors and cobblers.

> [Preface to *Uber die spezielle und die allgemeine Relativitatstheorie* quoted in: L.K. Nash (1968) *Elements Of Statistical Thermodynamics* Addison-Wesley, Reading MA, p. *v*.]

This type of elegance is a wonderful addition to any abstraction, but it is a secondary consideration.

On the other hand, if 'elegance' means 'depth' and 'insightful correlation to the real world', then elegance does indeed become a primary consideration. A scientist begins with the freedom to parcel the real world into items and interrelations in any way that he wishes – at first blush, the world is his oyster. Nonetheless, which particular items of the real world he chooses to map onto the elements of his abstraction and which particular relations in the real world he chooses as the basis of the topology of his abstraction, these choices will determine whether his scientific abstractions are merely utilitarian or are, instead, truly revelatory. Again, I quote Einstein:

The set of formulae, which for you is all there is to an explanation, has to be consistent with the philosophy of nature in order to be a true explanation. Otherwise it is only a convenient device for predicting the future of a system, but does not give a real insight into its nature.

> [C.G. Fernandez (1956) 'My tilt with Albert Einstein' *Amer. Scientist* **44**, 208.]

How exactly does one build this type of elegance into a scientific abstraction? How does one best carve up the real world, choosing the proper real world items and the proper real world relations? In part, making these choices is one of the arts of science. Science is a human creation: the scientist actively builds the constructure of his science; he does not passively copy the world. Although there are better and worse partitionings of the real world, there are no 'right' partitionings, and the process of partitioning the real world into items and into topological relations is more an art than a precise science. The natural scientist can operate quite effectively and can, in fact, carry out serious science without justifying the absolute, true or independent existence of the patterns that he has chosen to study. For this reason, the scientist can boldly decree the definition of certain categories of patterns – if he can argue that these categories are scientifically useful – without having to prove that these categories really exist in some absolute sense outside of the completely subjective realm.

> All concepts, even those which are closest to experience, are from the point of view of logic freely chosen conventions...
> > [A. Einstein (1949) 'Autobiographical Notes', *Albert Einstein: Philosopher–Scientist* (P.A. Schlipp, ed.) Open Court, La Salle IL, p.33.]

There is, however, at least one important hallmark for identifying those items in the real world that can give scientific abstractions the type of elegance that I would call 'depth'. This hallmark is that the 'proper' real world items can be independently modelled as more detailed scientific abstractions.

> actual scientific research has thus far shown the need to analyse nature in terms of a series of concepts that involve the recognition of the existence of more and more kinds of things...
> > [D. Bohm (1971) *Causality And Chance In Modern Physics* Univ. of Pennsylvania Press, Philadelphia, p. 140.]

Deeper, more detailed scientific abstractions are built from the parts of the items of their parent abstractions, and the more detailed scientific abstraction must then still match appropriately with the parent scientific abstraction which has, in turn, been built from the whole items themselves. 'Proper' real world items have parts, so that each item itself can be modelled as a miniature scientific abstraction.

Deep scientific abstractions

A scientific abstraction has depth when one can look within it and discover still more detailed abstractions. A deep scientific abstraction is built from other internal scientific abstractions. Operationally, this means that a scientist has been able to split the original real world items into more parts, to independently build a new abstraction from these parts, and then to successfully match the two abstractions. This is not always an easy task. As in the construction of any abstraction, the scientist first accumulates a set of candidate parts – the bolts and nuts and springs – and he then decides which particular parts and which particular relations between those parts to include in his model. When further dissecting real world items, however, he always faces the problem that some of the parts of the real world items may not be apparent – he may not be able to discern or to recognize the internal clockwork.

Difficult as these tasks may be, the inescapable requisite for any deep scientific abstraction is that it be matched to items in the real world that have parts. Otherwise, the abstraction is a blind alley – we need parts to work our explanatory magic:

We might... argue that the phenomena of Nature are simple, and that it is we who introduce the complication, forced to do so by the limitations of our power of mind and our mathematics, which do not allow us to grasp the sublime simplicity of Nature otherwise than by the devious route of first splitting it up into partial aspects, so chosen that we can master them by our mathematics, and then recombining these to an approximate representation of the whole. However this may be, we cannot but use the only means at our disposal, and if the truth be one and simple, we can only comprehend it as a synthesis of many constituent truths.

> [W. de Sitter (1928) 'On the rotation of the earth and astronomical time' *Nature* **121**: 99.]

The separation of the putative physical world and our human percepts of it [e.g., H. Weyl (1949) *Philosophy Of Mathematics And Natural Science* Princeton Univ. Press, Princeton] has led some philosophers to insist on the continued divisibility of scientific abstractions [D. Bohm (1971) *Causality And Chance In Modern Physics* Univ. of Pennsylvania Press, Philadelphia]. This appears to underlie Bertrand Russell's epistemology:

as to what the events are that compose the physical world, they are, in the first place, percepts, and then whatever can be inferred from percepts... But on various inferential grounds we are led to the view that a percept in which we cannot perceive a structure nevertheless often has a structure...

> [R.E. Egner & L.E. Denonn (eds.) (1961) 'Physics and neutral monism', *The Basic Writings Of Bertrand Russell* Simon & Schuster, NY, p. 610.]

Perhaps this is one reason why Einstein found it hard to accept quantum mechanics as a fully satisfying scientific abstraction. Quantum mechanics is built upon real world items best represented by probability distributions with no discrete parts. In quantum theory, there is no possible reduction of the essential real world items, because these items exist only as whole continua – parts of such continua have no meaning. Immensely broad as it is in scope, there is a sense in which quantum theory is not deep. Einstein hoped for a deeper level, more complete theoretical explanation. 'Some physicists, among them myself,' wrote Einstein [(1940) 'Considerations concerning the fundaments of theoretical physics' *Science* **91**: 487–92], 'can not believe that we must abandon, actually and forever, the idea of direct representation of physical reality in space and time; or that we must accept the view that events in nature are analogous to a game of chance.' And, 'I incline to the belief that physicists will not be permanently satisfied with... an indirect description of Reality' [quoted in: A. Pais (1982) *'Subtle is the Lord...' The Science And The Life Of Albert Einstein* Oxford Univ. Press, NY, p. 463].

2

The nature of explanation

'What do you mean by that?' said the Caterpillar, sternly. 'Explain yourself!'

'I can't explain myself, I'm afraid, Sir,' said Alice, 'because I'm not myself, you see.'

(*Alice's Adventures In Wonderland*, Chapter V.)

Scientific abstractions answer the question: 'What?' and, more specifically, they answer the question: 'What is the exact form of pattern x in the real world?' Scientific abstractions formally describe the elements and the relations in certain real world patterns; they do so in a manner that maps the abstract realm directly back to the real world; and, in broad terms, this is the meaning of 'what something is'.

To be precise, however, there is a critical caveat to this simple characterization. Scientific abstractions are called 'abstractions' because they abstract certain features from the real world. Abstractions are *selected* representations; they are not full representations, and they are not the real world items themselves. These are important distinctions, for the only truly complete answer to 'What is the form of pattern x in the real world?' is a display of the actual pattern itself [J. Bronowski (1978) *The Origins Of Knowledge And Imagination* Yale Univ. Press, New Haven]. Any other answer, any rephrasing of pattern x, any reproduction, any metaphor – i.e., any abstraction – implies that we have been at work selectively viewing and, in fact, selectively structuring the real world pattern, and this is unavoidable.

Our morning eyes describe a different world than do our afternoon eyes, and surely our wearied evening eyes can report only a weary evening world
[J. Steinbeck (1962) *Travels With Charley.*]

The unavoidable intervention of the scientist between the real world and his representation of the world means that a scientific abstraction is not a full answer to the question: 'What?' nor even to the question: 'What is the exact form of pattern x in the real world?' Instead, if we are being precise, scientific abstractions answer: 'What exact form shall we explicitly propose for pattern x in the real world?'

This key qualification notwithstanding, I will summarize the role of scientific abstractions simply as answers to 'what' questions. This simple definition puts scientific abstractions in clear contrast to scientific explanations. For, in equally simple terms, the role of a scientific explanation is to answer 'how' questions.

Scientific explanations are determinate answers to 'how' questions

At first glance, this definition may seem to differ from the usual meaning of the word 'explanation'. An explanation is commonly thought of as the answer to a 'why' question: Why is the sky blue? Why is $1 + 1 + 1 + 1 + 1$ equal to 5? Why is the body of a person organized as it is? Such 'why' questions include 'how is it that?', but they also include more teleological undercurrents, such as 'what is the meaning of?' and 'what is the purpose of?'. Purpose and meaning encroach on the subjective and the spiritual, and full-blown 'why' questions can lead beyond the rather constrained realm of natural science.

A scientific explanation is but a modest form of explanations in general, skirting the subjective meaning of phenomena. Rather than being answers to 'why' questions, scientific explanations are answers to the more confined 'how' questions: How is it that the sky is blue? How is it that $1 + 1 + 1 + 1 + 1$ is equal to 5? How is it that a body of a person comes to be organized as it is?

The switch from 'why' explanations to 'how' explanations, a striking part of Galileo's science, marked the beginning of modern scientific explanations.

We should realize Galileo's position. He had few forerunners, a few friends sharing his views, but was strenuously opposed by the dominating philosophical school, the Aristotelians. These Aristotelians asked, 'Why do the bodies fall?' and were satisfied with some shallow, almost purely verbal explanation. Galileo asked, 'How do the bodies fall?' and tried to find an answer from experiment, and a precise answer, expressible in numbers and mathematical concepts. This substitution of 'How' for 'Why', the search for an answer by experiment, and the search for a mathematical law condensing experimental facts are commonplace in modern science, but they were revolutionary innovations in Galileo's time.

[G. Polya (1954) *Mathematics And Plausible Reasoning* Princeton Univ. Press, Princeton, Vol. I, pp. 194–5.]

What precise and standard form should 'how' questions take? To begin, a scientific explanation should be built of abstractions. In this way, the scientist can work with it at his desk, in the abstract and apart from the real world. Moreover, the abstractions should be formal – they should be scientific abstractions – so that the explanation will have a precise and internally consistent structure and so that the explanation can be directly related to the real world. In other words, a scientific explanation should be portable and it should be composed of well-defined elements connected in an explicitly specified structure.

By themselves, scientific abstractions are not explanations. An abstraction can only teach us 'what', while an explanation goes beyond the 'what' of a thing and gives us some insight into its 'how', particularly, into how that thing has come about. The 'how' of a scientific explanation implies that there is some action, some operation, that is the basis for the 'what' that is to be explained, and the essential and central importance of action in a scientific explanation has been thoroughly argued by P.W. Bridgman [e.g., (1936) *The Nature Of Physical Theory* Princeton Univ. Press, Princeton] in his 'operational' description of the synthetic sciences. Simply, a scientific explanation can be considered to be an operation that explicitly transforms a precursor abstraction into the final abstraction, the thing to be explained. In fact, the question: 'How is it that pattern x comes about?' might be more specifically rephrased as: 'Given that pattern x exists, what is its predecessor and by what operation is this predecessor transformed into x?' A scientific explanation is a formal answer to this question.

A scientific explanation might be defined as follows:

A scientific explanation comprises two different scientific abstractions, a final abstraction and a precursor abstraction, both of which have been independently generated from the same pattern. These two abstractions are linked together by a well-defined determinate operation through which the precursor abstraction can be transformed into the final abstraction.

Symbolically

$$\text{precursor abstraction} \xrightarrow{\;Op\;} \text{final abstraction}$$

Here the transforming operation, Op, should be determinate: there is no room for ambiguity in the rule for matching the elements and the connections of the precursor abstraction to the elements and the connections of the final abstraction. Sufficient information should be contained within a scientific explanation to decide definitely how to transform any particular precursor. We are not allowed to say 'maybe' or 'sometimes' or even '50 per cent of the time' – each precursor abstraction must become a specific, definite, and unfuzzy final abstraction.

Why determinate? It is because humans appear to be most comfortable with determinate explanations, and the goal of a scientist should be to build a human explanation. At base the human mind understands determinately – when stripped of its fancy trappings the human understanding of any phenomenon is fundamentally that some definite cause unfailingly precedes a single particular effect. The continued attempt to frame all explanations in such determinate terms and the rankling irritation when direct cause and effect scenarios cannot be formulated – these provide the stimuli driving modern synthetic science.

It is surely the search for a causal difference behind every difference in phenomena which has prompted every discovery.

> [K.J.W. Craik (1967) *The Nature Of Explanation* Cambridge University Press, p. 39.]

And, as Popper wrote:

And however successfully we might operate with probability estimates, we must not conclude that the search for precision laws ['determinate law'] is in vain.

> [K.R. Popper (1959) *The Logic Of Scientific Discovery* Basic Books, NY, p. 247.]

This goal poses quite a challenge. Determinate models are only an approximation to the real world; the fundamental nature of subatomic particles, for example, appears to be indeterminate. Thus, science is faced with the task of carving the real world into patterns for which determinism is, in fact, a good approximation. Challenging as it is, this task is not impossible – at some level, even stochastic processes can be written in terms of determinate scientific explanations. (See Chapters 7 and 8 and the Appendix.)

I belabor the importance of determinate explanations because of the epistemological dilemma at the center of twentieth century physics. In the face of the traditional acceptance of determinate explanations and counter to the tacit assumption that this determinism reflects the true behavior of natural phenomena, experimental evidence now suggests that certain fundamental physical phenomena are inherently indeterminate, and today's scientific explanations of fundamental physical phenomena explicitly mirror this indeterminism.

Nonetheless, as K.J.W. Craik [(1967) *op. cit.* p. 39] and D. Bohm [(1971) *Causality And Chance In Modern Physics* Univ. of Pennsylvania Press, Philadelphia] have argued, human science is best served by determinate or strictly causal explanations. The human mind needs explanations with a determinate cast, and we intuitively and automatically reshape even the most indeterminate formulations into cleaner and crisper structures. As an operational metaphysics, we naturally assume a determinative world, and most of contemporary science works, at least on a day to day basis, with such implicit beliefs as:

There can be no doubt that probabilities, as they are known to us, are creations of the human mind. An omniscient being who knows all the mechanisms of the universe in all details would need no probabilities.

[E. Borel (1965) *Elements Of The Theory Of Probability* (trans.: J.E. Freund) Prentice-Hall, Englewood Cliffs, NJ, p. 165.]

Determinate epistemologies are natural to the human mind.

Truly indeterminate explanations lack depth, and it can be difficult, if not impossible, to further reduce them to predecessor explanations. In the human world, the most useful scientific explanations are deep and determinate and whenever possible a scientist should strive to create determinate explanations. The peculiarities of quantum phenomena should not dictate that all scientific explanations must be indeterminate at base. Similar concerns undoubtedly motivated Einstein to write:

I cannot make a case for my attitude in physics which you would consider at all reasonable. I admit, of course, that there is a considerable amount of validity in the statistical approach which you were the first to recognise clearly as necessary given the framework of the existing formalism. I cannot seriously believe in it because the theory cannot be reconciled with the idea that physics should represent a reality in time and space, free from spooky actions at a distance. ... I am quite convinced that someone will eventually come up with a theory whose objects, connected by laws, are not probabilities but considered to be facts, as used to be taken for granted until quite recently. I cannot, however, base this conviction on logical reasons, but can only produce my little finger as witness, that is, I offer no authority which would be able to command any kind of respect outside of my own hand.

(Written to Max Born, *The Born–Einstein Letters*, no. 84.)

Predictions

The formal language built into a scientific explanation forces an important crispness onto the scientist's thoughts because formality demands precision. A good scientific explanation can be written as a computer program, in which colorful and weighted words are not acceptable, half thoughts must be completed, vague relations must be made concrete.

In addition, once a precise scientific explanation has been assembled, it can be dissociated from the real world. In the abstract realm, it can then be rotated about and seen from various angles, perhaps angles that are normally inaccessible or even unnatural, and it can be manipulated, modified, or dismantled in any variety of abstract maneuvers or thought experiments. In this way, scientific explanations become the scientist's infinitely tinkerable pocket models.

The essential contribution of scientific explanations, however, is their role as the constructure of science. Scientific explanations tie together disparate scientific observations and interrelate individual bits of data. The fabric and the scaffolding of science are its explanations, and the synthetic perspectives of science are entirely dependent on an underlying coherent network of scientific explanations.

These three attributes describe the fundamental uses of scientific explanations: *precision, tinkerability,* and *synthesis.* In the office and the laboratory and the field, they each appear in innumerable guises; but the most direct and telling application of scientific explanations, the most well-worn and important paradigm, is the prediction of new features of the real world. Given one set of observations, a good scientific explanation should logically lead to other observations.

In their roles as prognosticators, scientific explanations partake of their three key characteristics: they are synthetic (they logically tie together independent scientific observations), they are tinkerable (they are manipulated in the abstract realm to generate the predictions), and prediction from a scientific explanation is precise (it should be largely programmable and it should follow the logic of demonstrative reasoning). Predictive scientific explanations populate the theoretical sciences, and the advance that theory offers over empiricism is a precise, organized and abstract methodology for making predictions.

From a determinate scientific explanation, we can generate specific predictions. Although the particular predictions that we propose may turn out to be incorrect, our scientific explanation has proved useful if it has led to testable hypotheses about the real world, and scientific explanations naturally generate testable hypotheses. These hypotheses come from thought experiments – experiments that are carried out entirely in the abstract realm – but, because the precursor abstraction and the final abstraction of the explanation have each been isomorphically mapped to real world patterns, the scientist can directly relate individual variants of his abstractions back to the real world. He can repeatedly test new versions of his abstractions, and, even when these new versions depart radically from his original abstractions, the scientist can regain his real world bearings by patiently retracing his steps back to the original abstractions.

Operationally, the predictions from a scientific explanation are hypothetical variants of the pattern upon which the explanation was initially built. Such variants are constructed in two steps:

1 Modify the precursor abstraction.
2 Apply the transforming operation.

The outcome will be a modified final abstraction, and this new abstraction is now a prediction about the real world. If our scientific explanation is a good approximation to the real world, then similar changes in the real world precursors should produce the predicted changes in the final real world pattern. Is this actually the case for our particular pattern? After ruminating and wrestling in the abstract realm, we need only venture out into the real world and see.

Explaining the human figure

A rhetorical, narrative analysis of metaphysical issues can easily slip into artful featherstitching. To some extent, this danger can be reduced by careful logic; nevertheless, even when starting from a firm base and proceeding by careful steps, the whole artifice may gradually become eccentric until it is simply too far from the essential issues and too tenuously tied to the real world to be of much use. There is no absolute protection against this danger, but there is one procedure that definitely provides some balance: the argument must continually be reexamined in terms of specific examples.

As a specific example, let me explain the human figure. From our working definition, we know that such an explanation should take the form:

$$\text{precursor abstraction} \xrightarrow{Op} \text{final abstraction}$$

where the precursor abstraction has been independently derived from the same pattern (here, the human figure) as has the final abstraction.

We already have a final abstraction, the standard stick figure (Fig. 1). At the moment, we have few rules for choosing precursor abstractions, so let me arbitrarily build the predecessor from a more detailed examination of the human body. For example, the elements of our precursor abstraction might be: stick fingers, stick palms, stick arms, stick legs, stick feet, stick trunk, stick head, stick eyes, stick ears, stick nose and stick mouth – all connected in the form of a detailed stick figure with the topology of the human body. Likewise, we have not yet proposed many restrictions for the transforming operation, so I will arbitrarily use:

Replace the stick fingers and stick palm of each side of the body with stick hands. If found alone the stick palm is left unchanged. If found alone the stick fingers are ignored. Ignore the stick foot. Leave all other body parts and interconnections unchanged.

This is a useful scientific explanation because it leads to testable predictions. Changes in the precursor abstraction produce changes in the final abstraction. For example, if the stick fingers are deleted from the predecessor, then the transforming operation produces a final pattern with only a stick palm instead of an entire stick hand at the end of its stick arm. Therefore, if this particular scientific explanation is a good model of the real world, we can make the following prediction: when the finger precursors are specifically deleted in a human embryo, only a palm will develop. Experimental embryology shows that this is, in fact, the case.

Standardized scientific explanations

To standardize scientific explanations, the range of precursor abstractions and the variety of transforming operations must both be restricted. One natural and elemental class of precursor abstractions contains those abstractions that portray all of the possible interactive potentials of the precursor elements. These abstractions – the local precursor abstractions – encompass all potential interconnections among the raw materials, and they represent, in a sense, the 'universal laws' governing the assembly of patterns with these materials. Local precursor abstractions are incognizant of any one particular global topology; instead, they see an arbitrarily-ordered collection of items, among which all possible neighbor–neighbor relations can potentially be instituted. Local precursor abstractions have a local view; they blindly embody all possible patterns that can be built of the raw materials at hand. The local precursor abstractions are like children, they want to do everything at once, and it is only through scientific explanations that they mature into one particular 'adult' pattern. Scientific explanations based on local precursor abstractions will be the standard tool for arguments throughout the remainder of this book.

For instance, to explain the human figure from a local precursor abstraction, we begin with all of the raw materials needed for the final stick figure: stick hands, stick arms, stick legs, stick trunk, stick head, stick eyes, stick ears, stick nose and stick mouth. To construct the local precursor abstraction itself, the linkages between these elements are determined only by the 'universal laws' governing their potential interrelations, not by the actual interrelations that they happen to manifest in any one particular pattern. Here, the appropriate interrelation is physical attachment – the stick ear is connected to the stick head in the final abstraction because the human ear is physically attached to the human head in the real world – and, when constructing connections between two elements in the local precursor abstraction, the question becomes: 'If given the opportunity, could the real world items represented by these elements develop a physical attachment?'

Imagine that we have accumulated sufficient biological observations to argue that any two body parts could potentially develop a physical attachment if given the appropriate opportunity. Our local precursor abstraction is now a 'stick figure' in which all of the parts are interconnected, and, unless we take the additional liberty of highlighting certain links, the human figure can no longer be seen in this abstraction. The human figure is embedded somewhere in the precursor abstraction but so are a great many monstrosities.

In creating our full explanation, we began with a final abstraction, the familiar stick figure, and now we also have a local precursor abstraction. To formulate a standard scientific explanation, our final task is to devise a transforming operation that will change the local precursor abstraction into the final abstraction. Both abstractions contain identical elements. Therefore, the rule for transforming the *content* – the individual elements – is simply the identity operation: 'leave the elements unchanged'.

An appropriate rule for transforming the *configuration* – the individual linkages – is, however, more complex. One straightforward rule explicitly lists each of the linkages to be conserved during the transformation from precursor abstraction to final abstraction. For example: 'retain the connection between the stick ears and the stick head, retain the connection between the stick nose and the stick head, retain...' For detailed stick figures, this rule is quite lengthy. Is there a shorter rule? Can one invent a transforming operation that does not entail the explicit specification of each and every individual interrelation in the pattern of interest? Here, it is perhaps surprising that the answer is 'no': we have constructed a pattern-assembly system for which there is no smaller and simpler explanation.

In the natural world, there are many pattern-assembly systems for which there is no simple explanation. There are useful scientific explanations for these complex systems, but the final patterns that they produce are so heterogeneous that they cannot effectively be reduced to smaller or less intricate predecessor components. As I will argue in Chapters 7 and 8, these patterns are, in a fundamental sense, irreducibly complex, and our particular model of the generation of the human figure is an example of a pattern-assembly system that is irreducibly complex.

A formalized scientific explanation

To explain something is to give insight into how it is that this thing has come about, and scientific explanations are reductionist: they explain by reducing a phenomenon to its component parts. The full explanation describes the component parts and also describes those interactions between the components that actually conspire to bring about the whole phenomenon.

In a reductionist explanation, the scientist translates certain real world phenomena into other terms, terms that he knows and understands independent of the phenomena being explained [E. Nagel (1961) *The Structure Of Science. Problems In The Logic Of Scientific Explanation* Harcourt, Brace & World, NY]. He works his explanatory alchemy by transmuting plants and planets into pointer readings and by transforming neutrons and nebulae into numbers. Formally, the steps in this reduction are [P. Duhem (1954) *The Aim And Structure Of Physical Theory* Princeton Univ. Press]:

(a) Demarcate natural patterns.
(b) Split the patterns into component parts.
(c) Model the patterns as scientific abstractions composed of the component parts.
(d) Build a scientific explanation.

Having followed this recipe, the scientist can then apply a practical test of the formality of his abstract constructions. This test is the extent to which his constructions can be automated – i.e., written as a completely specified algorithm or computer program. The fewer the number of places in the algorithm that must begin 'and then a scientist does...', the more formal the construction can be considered to be.

Absolutely objective and complete formalizations cannot be achieved in any abstract realm. Nonetheless, a careful formalization remains indispensible for securely and coherently interweaving complex logical arguments. By translating narrative logic into mathematics, the scientist opens new structural vistas in the arguments, and he can identify arcane characteristics and obscure connections. Although it is destined to be imperfect, a formalized logical argument is still the grand goal.

Beyond translating his scientific explanations into an algorithm the scientist has further goals in his formalization. One of these goals is to create a common, standard, mathematical form so that even disparate phenomena can be compared. Local scientific explanations can be nicely captured in a simple standard mathematical form. (See the Appendix for fuller details.) Each abstraction – the precursor abstraction and the final abstraction – is represented as a graph. The elements of the abstraction – the raw materials – are the vertices of the graph and the interconnections between elements are the edges of the graph. (See Fig. 3, in the Appendix.) The local precursor abstraction is a graph in which all possible links are manifest, whereas the final abstraction is a graph that mirrors one particular pattern and in which only certain of the links exist. The transforming operation, then, eliminates those potential links in the precursor graph that do not actually appear in the final graph.

A graph can be summarized as a binary adjacency matrix, M, e.g.:

$$M = \begin{array}{c} \\ a \\ b \\ c \end{array} \begin{array}{ccc} a & b & c \\ 0 & 1 & 0 \\ 1 & 0 & 1 \\ 0 & 1 & 0 \end{array}$$

where a one in the (i,j) position indicates that elements i and j are connected and a zero indicates that they are not connected. In this way, a local scientific explanation:

$$\text{local precursor abstraction} \xrightarrow{Op} \text{final abstraction}$$

can be written as the transformation of two square binary matrices, MM and AM, of the same size:

$$MM \xrightarrow{Op} AM$$

The transforming operation Op can be made into the combination of a simple matrix operation – logical addition $\wedge$ – and another explicit square binary matrix, T, and thus the complete formalization of a local scientific explanation is:

$$MM \wedge T = AM$$

where: MM is the 'matching matrix' that summarizes the local precursor abstraction, AM is the 'adjacency matrix' that summarizes the final pattern of interest, and T is the 'templet' matrix that summarizes the global topological constraints that must be applied to the raw materials in order to produce the final pattern.

By limiting scientific explanations to those containing local precursor abstractions, we explicitly focus on the topology or configuration of the pattern to be explained. In local scientific explanations, both the precursor abstraction and the final abstraction contain the same set of elements and the transforming operation for elements is the identity operation. Content is conserved.

Although the precursor abstraction and the final abstraction cannot differ in their content, they can differ in their topologies – i.e., the elements of one abstraction can be connected in a different fashion from the elements in the other abstraction. The connections in the final abstraction must directly mirror the topology of the particular real world pattern of interest. On the other hand, in local scientific explanations, the connections in the precursor abstraction are not limited to the topology of any one particular pattern. Instead, they embody all of the potential interactions between elements, regardless of whether those potentials are actually manifest in any specific pattern. If the elements are a group of magnets, then they would all have the potential to be physically attached and the precursor abstraction would be entirely interlinked. On the other hand, in any particular situation only those magnets that have the appropriate opportunities – i.e., those that are sufficiently close together – will actually become attached, and the final abstraction may take on any of a variety of forms, such as a chain or a circle of magnets.

All of the elements and the connections in a final abstraction are inherited from corresponding elements and connections in the local precursor abstraction. In local scientific explanations, the final abstraction is fully embedded in the precursor abstraction, and it is exactly in this sense that a local precursor abstraction represents an explanatory predecessor of the final pattern.

Configurational explanations

To emphasize their special role, scientific explanations containing local precursor abstractions can be called 'configurational explanations'. This name indicates that the local scientific explanations are specifically explanations of the topology or configuration of a pattern.

For example, consider the pattern *abc*, represented by the abstraction:

$a–b–c$

composed of the three elements $\{abc\}$. The adjacency matrix for this abstraction is:

$$
AM = \begin{array}{c@{\quad}ccc}
 & a & b & c \\
a & 0 & 1 & 0 \\
b & 1 & 0 & 1 \\
c & 0 & 1 & 0
\end{array}
$$

The matching matrix, summarizing the local precursor abstraction, for this explanation depends on a detailed description of the nature of the elements. We must determine which elements can, if given the appropriate opportunities, link together. Let us assume we find that the elements can all interconnect in any fashion, that they are intrinsically topologically naive. Thus, we have a matching matrix that is essentially nonrestrictive:

$$
MM = \begin{array}{c@{\quad}ccc}
 & a & b & c \\
a & 0 & 1 & 1 \\
b & 1 & 0 & 1 \\
c & 1 & 1 & 0
\end{array}
$$

Finally, the transforming operation is the logical addition of the templet matrix, T, representing the specific opportunities and constraints that must be imposed on the precursor elements to produce the particular final pattern. T is given by:

$$
\begin{array}{cccc}
 & a & b & c \\
a & 0 & 1 & 0 \\
T = b & 1 & 0 & 1 \\
c & 0 & 1 & 0
\end{array}
$$

Symbolically, the entire configurational explanation is:

$$MM \wedge T = AM$$

or, explicitly:

$$
\begin{bmatrix} 0 & 1 & 1 \\ 1 & 0 & 1 \\ 1 & 1 & 0 \end{bmatrix}
\begin{bmatrix} 0 & 1 & 0 \\ 1 & 0 & 1 \\ 0 & 1 & 0 \end{bmatrix}
\wedge
\begin{bmatrix} 0 & 1 & 0 \\ 1 & 0 & 1 \\ 0 & 1 & 0 \end{bmatrix}
=
\begin{bmatrix} 0 & 1 & 0 \\ 1 & 0 & 1 \\ 0 & 1 & 0 \end{bmatrix}
$$

(See the Appendix for a more detailed mathematical exposition and for further examples.)

This general epistemological partitioning of the world has been repeatedly invoked for explanations of complex phenomena. From Aristotle in his *Natural Science* ('The substratum [precursor elements] which persists and the form [templet] that is imposed are jointly responsible for whatever is created') through K.R. Popper, who proposes that all scientific ('causal') explanations are composed of 'universal laws' [potentials of the precursor elements] and particular 'initial conditions' [templets] (1959 *The Logic Of Scientific Discovery* Basic Books, NY, pp. 59–60) to M. Polanyi (1968 'Life's irreducible structure' *Science* **160**: 1308–12) in his hierarchical analyses of biological patterns and H.H. Pattee (1979 'Compelmentarity *vs.* reduction as explanation of biological complexity' *Am. J. Physiol.* **236**: R241–6) in his functional analyses, natural philosophers have frequently emphasized the essential contrast between the universal potentials of the elements and the limitations imposed by the specific system at hand. For instance, F. Jacob [(1977) 'Evolution and tinkering' *Science* **196**: 1161–6] begins his evolutionary analyses with the basic split between the 'constraints' (precursor elements) and the 'history' (templet) of a complex pattern:

Complex objects are produced by evolutionary processes in which two factors are paramount: the constraints that at every level control the systems involved, and the historical circumstances that control the actual interactions between the systems.

Configurational explanations are explicit representations of this split, and they fulfil R. Carnap's important definition of a scientific explanation. Carnap [(1966) *Philosophical Foundations Of Physics* Basic Books, NY] has argued that scientific explanations of particular phenomena must always contain both a universal law and the specific application of the law that will produce the phenomenon to be explained. In a configurational explanation, the universal law is represented by the matching matrix that defines the inherent and 'universal' potentials of the elements, and the templet represents the specific constraints that go into producing the particular final pattern to be explained.

Configurational explanations require three critical interactions with the real world. First, the final abstraction should map in a one-to-one fashion onto the real world pattern of interest. To do this, the scientist must partition his pattern into a definite set of items and he must identify a critical binary relation between these items – i.e., he carves the world into natural interrelated units.

Second, the local precursor abstraction should summarize the potential for the items to share this identified binary relation, whether or not that relation actually appears in the particular pattern being explained. To build the appropriate precursor abstraction, either the scientist must examine the items and then deduce their potential interactions or he must test the potential interactions by giving the items sufficient opportunities to interact – i.e., he experiments with the natural units.

Third, the scientist should attempt to ascribe a real world meaning (a real world correlate) to the matrix T, to the 'templet', the protagonist of a configurational explanation. Frequently, the two other matrices, MM and AM, are directly determined from the real world. T, on the other hand, is often initially generated entirely in the abstract realm. Can the scientist make a real world sense of T? Successfully relating T to the real world provides one of the special insights offered by configurational explanations.

3

Configurational explanations

There is one ideal of survey which would look into each minute
compartment of space in turn to see what it may contain and so
make what it would regard as a complete inventory of the
world. But this misses any world-features which are not located
in minute compartments. We often think that when we have
completed our study of one we know all about two, because
'two' is 'one and one'. We forget that we have still to make a
study of 'and'. Secondary physics is the study of 'and' – that is
to say, of organisation.

[A.S. Eddington (1929) *The Nature Of The Physical World*
Macmillan, NY, pp. 103–4.]

In *A Child's Garden Of Verses*, Robert Louis Stevenson wrote:

The world is so full of a number of things,
I'm sure we should all be as happy as kings.

The world is certainly bountifully supplied with different things, and
while these all appear new and wonderful to a child, they have largely
become old and familiar to an adult. A scientist, on the other hand,
dreams of a balance between the child world and the adult world.
The scientist would like most things to be old and familiar while a
few things still remain new and wonderful.

The scientific pursuit is turning the new and wonderful into the
old and familiar: a scientist tries to make the world around him
understandable. At the same time, the scientist needs a constant
supply of problematic observations to explain – without novelties,
discrepancies, and inconsistencies he would be out of business. For
this reason, much scientific effort goes into ferreting out disparities
in nature, revealing new and incongruous observations, and
discovering surprises. Nonetheless, in a complete science, this analytic
process must always be tempered by an equally essential synthetic
effort.

The myriad things in the scientist's world are all patterns, and the
synthetic process of turning the new and wonderful into the old and

familiar is the scientific explanation of natural patterns. Explanation unfolds largely in the abstract realm. Here, real world patterns are modelled as scientific abstractions composed of certain elements (their content) interconnected in a specified organization (their configuration), and thus a scientist can pursue two broad classes of explanation. On the one hand, he can ask: 'How is it that the given pattern has come to have its particular content?' or, he can ask: 'How is it that the given pattern has come to have its particular configuration?' Both are legitimate scientific inquiries, but I will focus only on the latter question. Its answer is a *configurational explanation.*

Templets

At the center of a configurational explanation, between the precursor abstraction and the final abstraction, separating the 'universal laws' of assembly from the one particular pattern that has actually been constructed, sits the pattern's templet. The templet is the embodiment of the particular organizational information, beyond that inherent in the elements of the precursor abstraction, needed to assemble the specific pattern represented by the final abstraction. A templet is the global topology that must be superimposed on the local topological potentials of the input in order to produce the particular pattern of the output. A templet is a necessary prepattern.

In common usage, a templet is a model, a prototype, a guide, a prepattern or a blueprint. Today, the word 'templet' is considered an archaic variant of 'template'. 'Templet' is actually the older of the two words, and 'template' appears to be an Anglicized alteration of 'templet', the original term. 'Templet' is probably a diminutive form of the French word *temple*, the wooden claw-like spreader used on a loom to keep the warp from being pulled out of shape at the edges. In turn, the French *temple* came from the Latin *templum*, meaning a crossbeam running along the roof and supporting the common rafters – perhaps the temple in a loom was reminiscent of the templum in a roof.

I have reintroduced the old word 'templet' because the modern word 'template' is now firmly associated with one particular natural templeting process, the translation and transcription of DNA. With the historic sentence:

We should like to propose... that each of our complementary DNA chains serves as a template or mould for the formation onto itself of a new companion chain.

[J.D. Watson & F. Crick (1953) 'The structure of DNA' *Cold Spring Harbor Symp. Quant. Biol.* **18**: 123–31.]

James Watson and Francis Crick irrevocably colored the scientific meaning of the word 'template'.

Unique templets

A configurational explanation is a formal scientific model for how the organization of a particular pattern has come about, cleanly breaking the causes of the pattern's topology into two distinct parts – the information inherent in the precursors and the additional configurational information in the templet. The templet is the sum of the effective opportunities and the effective external limitations actually available for connections between the raw material elements.

Sometimes, the templet is unique, and there is only one possible templet for building a particular pattern. Recall that a configurational explanation is symbolized as:

$$MM \wedge T = AM$$

In a configurational explanation, two square binary matrices are added with the logical operation $\wedge$, which has the addition table:

$\wedge$	0	1
0	0	0
1	0	1

Thus, for a templet to be unique – for the templet matrix to be the completely determinative agent in the matrix equation $MM \wedge T = AM$ – the matching matrix must contain only ones. A matching matrix filled with ones summarizes a totally malleable set of precursor elements, and a configurational explanation containing a precursor abstraction that permits interconnections between all of the elements will require a unique templet. From such a completely nonrestrictive precursor abstraction, the raw material elements can potentially form any configuration and thus they are topologically naive.

Topological naiveté is a property of many raw materials. Letters, for instance, can be arranged in almost any conceivable pattern, and it is for this reason that a monkey with a typewriter can print all possible essays. Eventually, the monkey will write a Shakespearean sonnet, but along the way he will also type a great deal of nonsense. To ensure that he actually prints a poem, a templet is needed, and, to ensure that he prints *Macbeth*, there is only one appropriate blueprint – the templet is unique.

Equivalent templets

On the other hand, the local topological information in the raw materials will sometimes restrict the potential configurations permitted among the given elements; in these cases, more than one templet can be used to construct any one particular output pattern, and a number of templets will be effectively interchangeable and equivalent.

It is easy to calculate the number of equivalent templets for any configurational explanation. Each zero in the matching matrix will necessarily lead to a zero in the adjacency matrix, regardless of the particular value that is added by the templet – the corresponding slots in the templet matrix are invisible. To compute the number of distinct but equivalent templets, count the number N of zeros in the matching matrix. These zeros in the matching matrix represent slots that can correspond to either a one or a zero in the templet martrix, so the number of equivalent templets is 2^N.

When the raw materials are topologically naive, the matching matrix is entirely filled with ones, $N = 0$, and there is only one equivalent templet. As N increases, the number of equivalent templets increases exponentially: the uniqueness of the templet is inversely related to the number of potential interconnections among the precursors. In a configurational explanation, there is a balance between the specificity of the local topological information and the specificity of the global topological information: when the precursors are topologically naive, the templet is explicit and unique, but when the precursors are topologically knowledgeable, the templets need not be as specific. When the 'universal assembly laws' permit any configuration, particular configurations must be determined by a detailed set of specific situational specifications, but when the 'universal assembly laws' dictate particular configurations, few additional constraints need be imposed.

Maximal templets

No matter how knowledgeable are the precursors and how restrictive are the underlying 'universal assembly laws', one can always use a completely specified templet – the binary matrix that is identical to AM, the adjacency matrix of the final abstraction. For example:

$$
\begin{bmatrix} 1 & 1 & 1 & 1 \\ 1 & 1 & 1 & 1 \\ 1 & 1 & 1 & 1 \\ 1 & 1 & 1 & 1 \end{bmatrix} \wedge
\begin{bmatrix} 0 & 1 & 0 & 0 \\ 1 & 0 & 1 & 0 \\ 0 & 1 & 0 & 1 \\ 0 & 0 & 1 & 0 \end{bmatrix} =
\begin{bmatrix} 0 & 1 & 0 & 0 \\ 1 & 0 & 1 & 0 \\ 0 & 1 & 0 & 1 \\ 0 & 0 & 1 & 0 \end{bmatrix}
$$

as well as:

$$
\begin{bmatrix} 0 & 1 & 0 & 0 \\ 1 & 0 & 1 & 0 \\ 0 & 1 & 0 & 1 \\ 0 & 0 & 1 & 0 \end{bmatrix} \wedge
\begin{bmatrix} 0 & 1 & 0 & 0 \\ 1 & 0 & 1 & 0 \\ 0 & 1 & 0 & 1 \\ 0 & 0 & 1 & 0 \end{bmatrix} =
\begin{bmatrix} 0 & 1 & 0 & 0 \\ 1 & 0 & 1 & 0 \\ 0 & 1 & 0 & 1 \\ 0 & 0 & 1 & 0 \end{bmatrix}
$$

although in the former case the templet is unique and in the latter case it is one of 1024 distinct but equivalent templets.

In general, it will always be true that:

$$MM \wedge AM = AM$$

AM is the maximal templet for the final abstraction; it is identical to the output pattern, and it provides the most detailed explicit specification of global topology that is ever needed to determinately build the pattern of interest. Although not always necessary, it is always possible to use a maximal templet, and, when it is used unnecessarily, the topology is simply overdetermined. Overdetermined explanations are sometimes good explanations for real world phenomena because Nature is not limited to operating in the most parsimonious manner, and topology is, in fact, overdetermined in the formation of many natural patterns.

S, the minimal necessary size of a templet

Maximal templets must be invoked to explain the configuration of certain patterns, but other patterns can be explained with smaller templets. These latter patterns are assembled from constituents that can only be configured into a few distinct patterns (the 'universal assembly laws' are quite restrictive) and only minimal additional topological information need be superimposed on that already embodied in the raw material elements themselves.

In the formalism of configurational explanations, all templets, be they maximal or minimal, are represented by a matrix of a standard physical size – the dimension of the adjacency matrix. On the other hand, there is some important sense in which the effective sizes of the templets can differ from their physical sizes, and it would be quite useful to have a simple measure of this effective size, the minimal necessary size of the templets for a given configurational explanation.

One obvious measure equates the minimal necessary size, S, of a templet with the number of slots in its matrix that must be explicitly specified as either a one or a zero. For a configurational explanation to operate determinately and to produce one and only one output pattern, each one in the matching matrix requires an explicitly specified slot in the templet matrix. Hence:

$$S = \sum_{\text{row}} MM$$

Given a final pattern and a set of precursors, S is the minimal necessary size for a templet that will complete the appropriate configurational explanation. A configurational explanation is built around a certain physical matrix size – the size of the adjacency matrix – and S tells how much of that matrix size must be specified with a particular global topology in order to produce the pattern to be explained.

S is measured in 'matrix slot' units, and, because all of the matrices are binary, 'matrix slots' are equivalent to the standard 'information bits'. A templet always embodies sufficient information to distinguish one particular configuration from all of the different configurations permitted by a particular matching matrix, and S is the number of bits of global topological information necessary to assemble a final pattern from a particular set of precursors.

If E is the number of equivalent templets for a particular configurational explanation and if S_{max} is the size of the maximal templet (the physical size of the matrices), then:

$$E = 2^{(S_{max} - S)}$$

E has an exponential inverse relation with S: as the number of equivalent templets increases, the minimal necessary size of the templets decreases. When the minimal necessary size of the templets is small, the number of equivalent templets is large; e.g.:

when $S = 0$, then $E = 2^{S_{max}}$

In contrast, when the minimal necassary size of the templets is large, the number of equivalent templets is small; e.g.:

when $S = S_{max}$, then $E = 1$

Thus, *unique* templets are truly *maximal* templets.

Programs of configurational explanation

although it is established that there are limitations to the powers of any particular machine, it has only been stated, without any sort of proof, that no such limitations apply to the human intellect. But I do not think this view can be dismissed quite so lightly.

[A.M. Turing (1956) 'Can a machine think?' *The World Of Mathematics* (ed. J.R. Newman) vol. 4, p. 2109.]

Machines

Humans cannot disentangle themselves from their creations, and this truism belies the platitude: 'machines will set man free'. Machines free our time in one realm and occupy our time in another. At heart, however, most of us do not begrudge machines the vast arenas that they occupy in our lives. Machines, as other creations, are extensions of ourselves; the depths of time and space that they plumb become depths that we can plumb, and the ornate designs that they form become reworkings of ourselves.

Machines are physical devices, but they have been very closely modelled in the abstract realm by a branch of discrete mathematics called Automata Theory. By 'machine', scientists usually mean a physical device, a gadget in the real world. Automata Theory deals with abstractions of machines, and one of the critical questions is always: 'Can the abstraction that I have described be realized as an actual machine?' To distinguish between the real world and the abstract realm, it is well to maintain this terminology: a machine is a physical device; an automaton is an abstraction, specifically, an abstract system that can exist in certain states and for which there are precise rules that describe how it changes its state.

It is important that configurational explanations be tied into Automata Theory for two reasons. First, practical experience has shown that Automata Theory provides a useful set of abstractions for a wide range of real world systems. Second, theoretical arguments demonstrate that Automata Theory can make explicit essentially all of the processes that underlie any computation.

Finite automata

The simplest automata are finite automata – 'finite' because they can exist in only a finite number of states. Together, a list of its possible states and the rules for transitions between states will fully define a particular finite automaton. To go farther and to describe the play enacted by a finite automaton, we must also specify its initial state and a sequence of rules – the instructions – that determine the specific state transitions that actually occur.

A configurational explanation can be transformed into a finite automaton by making some of its implied processes explicit, and this reveals its internal clockwork. Basically, a configurational finite automaton is a device that takes two inputs – a matching matrix and a templet matrix – and that produces one output – an adjacency matrix. Most automata use vectors (strings) instead of matrices, and by stringing out the matrices of our configurational explanation, we can create a configurational finite automaton that will operate with vectors. Each matrix can be turned into a vector by concatenating the rows, one after the other. Thus, the matching matrix:

$$
\begin{array}{c c c c}
 & a & b & c \\
a & 0 & 1 & 0 \\
MM = \quad b & 1 & 0 & 1 \\
c & 0 & 1 & 0
\end{array}
$$

becomes the matching vector:

$$MV = 0\ 1\ 0\ 1\ 0\ 1\ 0\ 1\ 0$$

Likewise, the templet matrix T can be transformed into a templet vector TV, and the adjacency matrix AM becomes an adjacency vector AV.

The entire configurational explanation is now the logical addition of two vectors.

$$MV \wedge TV = AV$$

and it is equivalent to a finite automaton – a LOGICAL ADDER,
LA – that will add any two binary vectors:

$$MV \longrightarrow \boxed{LA} \longrightarrow AV$$
$$TV \longrightarrow$$

There are no restrictions on the length or on the sequence of the
output vector, and there are no restrictions on the lengths or on the
sequences of the vectors to be input. Any conceivable binary vectors
can be input, and, because these vectors embody all of the relevant
configurational information, the automaton itself can be said to be
entirely naive as to the details of any particular pattern.

Such two-state finite automata are very simple devices, and the
logical adder has been well-studied and thoroughly exploited – it is,
for instance, one of the elemental cell types in the neural net
automatons, where it is called an 'AND cell'. A logical adder is
determinate, and in this way a final pattern can always be uniquely
predicted from its standard predecessors – the precursors and the
templet. On the other hand, logical adders are polarized: they cannot
be run *backwards* determinately, and specific predecessors cannot
always be uniquely predicted. Nontheless, the equivalence classes of
predecessors, and in particular the equivalence class of necessary
templets, can always be uniquely predicted from the other components of a logical adder. With the other two vectors, we can always
discover the key characteristics (such as the minimal necessary size
and the number of equivalent templets) of the necessary templets,
we can at least uniquely *categorize* the pattern predecessors.

Earlier, the process of understanding the topology of a pattern
was transformed into a set of matrix operations, a configurational
explanation, and, in so doing, the notions of 'pattern', 'topology',
and 'understanding' were given an explicit meaning. Now, configurational explanations have themselves been transformed into
particular finite automata (logical adders), thus making explicit
the internal operations – the machinery – within configurational
explanations.

Concatenation automata

One of the powerful techniques of scientific analysis is the abstraction of real world phenomena as isolated systems, and in most fields of science, segments of the world are regularly modelled as self-contained abstractions that can operate independent of outside influences. For example, classical thermodynamics is founded on the assumption that the systems under study can be treated as completely insulated from the rest of the world, likewise the Galilean abstraction partitions complex two-dimensional motion into two isolated one-dimensional components – a vector describing the motion in the x-direction and a vector describing the motion in the y-direction. The far-reaching utility of such independent partitionings demonstrates the fundamental importance of inventing ways to split the real world into independent units.

However

Nature never offers us simple and undivided phenomena... to observe, but always infinitely complex compounds of many different phenomena. Each simple phenomenon can be described mathematically in terms of the accepted fundamental laws of Nature, to interpret the complex phenomena of daily experience we must analyse them into their simple components, isolate these, and from each separately draw our conclusions.

[W. de Sitter (1928) 'On the rotation of the earth and astronomical time' *Nature* **121**: 99.]

The real world is not built of truly independent units; the complex systems of the real world are characterized by their widespread interdigitations with outside events, influences, and items. These interdigitations define the contexts of such systems.

Real world patterns, and most especially complex patterns, can be fully understood only in their rightful contexts, and the scientific analysis of a pattern should allow for precise statements about the external context of the pattern as well as about its internal content and configuration. By themselves, configurational explanations help to explain the internal structure of patterns, but the contextual understanding of patterns requires some additional statements – statements about assemblies of interrelated configurational explanations. A collection of configurational explanations that are concatenated in a well-defined manner might be called a 'program of configurational explanations', and certain programs of configurational explanations form scientific theories. As automata, configurational explanations are easily concatenated, and thus certain scientific theories can be directly modelled as concatenation automata.

For instance, logical adders alone can be concatenated in parallel; they can be stacked into theories formed of layered configurational explanations:

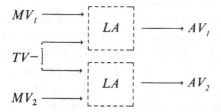

By themselves, logical adders can also be concatenated in series; they can be strung into theories that are chains of configurational explanations:

$$MV_1 \longrightarrow \boxed{LA} \qquad MV_2 \longrightarrow \boxed{LA} \longrightarrow AV$$
$$TV_1 \longrightarrow \qquad \longrightarrow TV_2 \longrightarrow$$

In parallel concatenation automata, the final patterns become standardized to a common templet, resulting in a matched output. This matched output is a selected, monad's view of a world:

... each created monad is a perpetual living mirror of the universe... every monad is a mirror of the universe after its manner...

G.W. Leibniz

[H.W. Carr (1930) *The Monadology Of Leibniz* Univ. California Press, LA, pp. 99 & 110.]

– the common templet in a set of parallel logical adders acts as an idiosyncratic crystal refracting and coloring the inputs that come into its ken, and the layered scientific theories built in this way explain patterns as selected highlights of particular templets.

Parallel adders can also be seen as decomposition automata for templets. Different matching vectors will select different portions of the templet vector to reproduce, and a set of parallel adders will produce a selected decomposition of a given templet. Here, we have reversed our perspective: the various matching vectors of a parallel adder now become the monads, and the templet becomes the variable input, the particular view of the universe. Each matching matrix refracts the templet through its particular crystal, and each output is a slightly different view of the same aspect of the world. When the matching matrices are the variables, the templet is the crystalline monad and the parallel adder is a filter; when the templet is the variable, the matching matrices are the monads and the parallel adder is a decomposition automaton.

A monadic vista characterizes parallel concatenation automata; on the other hand, a characteristic of serial concatenation automata is that their final patterns tend to become simpler (in the sense that they will contain fewer and fewer ones). Long strings of serial adders with a variety of matching vectors reduce the complexity of any input templets, and the chained scientific theories built in this way explain patterns as increasing simplifications of particular templets.

By interconnecting simple two-state automata it is possible to simulate any finite automaton, but in general you need additional elemental building blocks – one must build such simulations using other varieties of two-state automata in addition to the logical adder. To create a universal alphabet for scientific theories – to be able to construct any finite automaton – one needs two further elemental automata: disjunction automata (OR cells) and A AND NOT B automata. [For a good discussion of this issue, see M.L. Minsky (1967) *Computation: Finite And Infinite Machines* Prentice-Hall, chapter 3. See also the Appendix.]

Disjunction automata, DA, can be diagrammed as:

$$TV_1 \longrightarrow \boxed{DA} \longrightarrow TV_3$$
$$TV_2 \longrightarrow$$

or symbolized as:

$$TV_1 \lor TV_2 = TV_3$$

A AND NOT B automata, $A \sim B$, can be diagrammed as:

$$MV \longrightarrow \boxed{A \sim B} \longrightarrow AV$$
$$TV \longrightarrow$$

or symbolized as:

$$TV \wedge \sim MV = AV$$

[Using the symbolic form of the automata, concatenated automata can be directly analyzed with standard symbolic logic. For instance, the general rules for serial concatenations include:

$$(MV_1 \wedge TV) \wedge MV_2 = (MV_1 \wedge MV_2) \wedge TV$$
$$MV \wedge (TV_1 \vee TV_2) = (MV \wedge TV_1) \vee (MV \wedge TV_2)$$
$$(MV \wedge TV_1) \vee TV_2 = (MV \vee TV_2) \wedge (TV_1 \vee TV_2)$$

and the general rules for parallel concatenations include:

$$(MV \wedge TV_1) \vee (MV \wedge TV_2) = MV \wedge (TV_1 \vee TV_2)$$
$$(MV_1 \wedge TV) \vee (MV_2 \wedge TV) = (MV_1 \vee MV_2) \wedge TV$$
$$(MV \wedge TV_1) \vee (TV_2 \wedge TV_3) = ((MV \vee TV_2) \wedge (TV_1 \vee TV_2)) \vee TV_3$$
$$= (MV \vee TV_2 \vee TV_3) \wedge (TV_1 \vee TV_2 \vee TV_3)]$$

By incorporating all three elemental functions into concatenated automata, we gain the ability to build an almost limitless variety of complex functions, and we now put configurational explanations in quite wide, and essentially universal, contexts. With the cooperation of disjunction automata and *A* AND NOT *B* automata, configurational explanations can be set into the most comprehensive of scientific theories.

The advent of Automata Theory has led to an understanding of the inherent limitations of all machines, and this stimulated Turing to ask: 'Are there similar limitations to the human mind?' The representation of a scientific theory as a concatenation automaton sets certain human intellectual endeavors squarely in the realm of machines, and this maneuver hints that the answer to Turing's question might be: 'Yes.'

Nonetheless, the mechanization of so human a function as explanation does not immediately define the parallels between mind and machine. Configurational explanations, as any other epistemological devices, remain creations of the mind. Although they appear to obey general laws, such as the two principles of templeting (see Chapter 4), and although these laws define absolute constraints on the functioning of the epistemological devices, laws are ultimately determined by the particular epistemological split that we have decided to put upon the world from the very outset. Fundamental as it may be, that split is a figment of our imagination, and for this reason the mechanizations that we create continue to serve only at the pleasure of the mind. As J. Bronowski wrote:

Nature is not a gigantic formalizable system. In order to formalize it, we have to make the assumptions which cut out some parts. We then lose the total connectivity. And what we get is a superb metaphor, but it is not a system which can embrace the whole of nature.

[(1978) *The Origins Of Knowledge And Imagination* Yale Univ. Press, New Haven, p. 80.]

4

Templeting

I placed a jar in Tennessee,
And round it was, upon a hill.
It made the slovenly wilderness
Surround that hill.

The wilderness rose up to it,
And sprawled around, no longer wild.
The jar was round upon the ground
And tall and of a port in air.

It took dominion everywhere.
The jar was gray and bare.
It did not give of bird or bush,
Like nothing else in Tennessee.

(Wallace Stevens 'Anecdote of the jar'.)

What preexisting order must there be to make a pattern? How much specific information is needed, how much of a prepattern? In some sense, all of the configurational information of a pattern must already be present when the pattern is assembled, because one cannot create somthing from nothing. Just as determinate physics is based on the assumption that the stuff of the natural world cannot arise entirely *ex nihilo*, so determinate metaphysics is based on the assumption that the order of the natural world cannot materialize totally without precedent:

'concrete substances'...when they come into existence presuppose some kind of underlying substratum... For in every case there is something already present, out of which the resultant thing is born... The substratum which persists and the form that is imposed are jointly responsible for whatever is created.
[Aristotle *Natural Science.*]

Configurational information must be conserved, but the form of the information my change drastically in the process of putting together a pattern – the original forms can be transmuted, and new forms can be created. One can debate whether the whole is ever actually greater than the sum of its parts, but one can always point to examples where the whole is quite different from the sum of its parts.

Configurational explanations deal in the forms of the information of pattern assembly; they are explicit descriptions of the transformation of preexisting forms into final forms. Configurational explanations separate the causes – the preexisting forms – of pattern organization into two distinct camps: the inherent, 'universal' topological information in the precursor elements, and the additional external and situationally-specific topological information provided by templets. The distribution of configurational information between these two forms falls on a spectrum. At one end, most of the organizational information is embodied in the raw materials themselves (the precursor elements are topologically knowledgeable), only minimal templets are needed, and the patterns can be called 'self-assembling'. At the other end of the spectrum, the raw materials embody little restrictive organizational information (the precursor elements are topologically naive), an extensive templet is needed, and the patterns are largely templeted. Extensive templets provide more than just the opportunities for the precursor elements to self-assemble, they completely and explicitly specify the particular topologies produced by pattern-assembly processes. In a world where many patterns are possible, the assembly of one particular pattern necessitates an extensive templet.

By basing a metaphysics on clearly delineated templets and templeted patterns, two fundamental principles of pattern assembly immediately stand out. These principles are:

1 All patterns can be completely templeted.
2 Certain paterns must be completely templeted.

All patterns can be completely templeted

Regardless of whether the complete global topology *must* be explicitly specified when assembling a pattern, it always *can* be entirely specified. Consider a jigsaw puzzle that fits together into only one coherent pattern. A robot can be programmed to eventually create that pattern through a trial and error process: the 'universal assembly laws' inherent in the materials ensure an appropriate outcome, and the robot needs no explicit prepattern. On the other hand, if we wished, a complete blueprint of the final pattern can also be written into the computer program controlling the robot, and the final configuration of the finished puzzle will simply be overdetermined.

Formally, the first principle of templeting states that the adjacency matrix – the most explicit description of the final pattern – can always act as its own templet; i.e.:

$$MM \land AM = AM$$

This principle sets boundary values for configurational explanations. For $n \times n$ matrices, the minimal necessary size of AM, the maximal templet, is $S = S_{max} = n^2$, and thus each pattern needs no more than n^2 bits of global topological information for its construction. In addition to providing a standard size measure for any pattern, the first principle assures us that, given a pattern and its precursors, we can always explain the pattern – i.e., we can always produce at least one complete configurational explanation, a configurational explanation containing the maximal templet.

The first principle and Occam's razor

Configurational explanations cannot always be run backwards determinately: together, the 'universal assembly laws' for the precursor elements and the final pattern need not uniquely fix the particular templet that is actually operating in the real world. The natural world has no absolute requirement for minimization, especially in complex systems, and when the precursors are topologically knowledgeable, the actual templet may nonetheless be more extensive than is minimally necessary. Nature is not required to employ the smallest possible templet, and, when modelling real world phenomena, we cannot blindly invoke the most efficient configurational explanation.

The most parsimonious explanation is not necessarily the natural explanation – this idea runs against the usual scientific dogma; for instance, Newton wrote in his *Rules Of Reasoning In Philosophy*:

Rule 1

We are to admit no more causes of natural things than such as are both true and sufficient to explain their appearances. To this purpose the philosophers say that Nature does nothing in vain, and more is in vain

when less will serve; for Nature is pleased with simplicity, and affects not
the pomp of superfluous causes.

(*Philosophiae Naturalis Principia Mathematica.*)

A number of fundamental presumptions are mixed into this famous
statement of Occam's razor. Let me unravel two of the most basic,
beginning at the level of the explanation itself.

Parsimony should be the rule for scientific explanations

Explanations are human creations; they are our models of
and our approximations to real world phenomena, and they
are wholly denizens of the human abstract realm. As such, there are
no absolutely right abstractions or perfect explanations or true
theories – there are only better abstractions, more useful explanations, more encompassing theories. The goodness and the utility
of these creations is partly determined by the inherent nature of the
human nervous system – by our ways of perceiving and by the innate
logic of the human neural circuitry. At the same time, the goodness
and the utility of our metaphysical tools are also determined by
the metaphysical environment in which they are used, and this
environment has been woven from our cultural habits and from the
intellectual structures that we have culturally inherited.

Both of these agents – the structure of our brains and the
intellectual rules of our cultures – encourage parsimonious explanations. (See the appendix for a more detailed discussion of
parsimonious configurational explanations.) Even in the face of the
astronomical numbers of neural elements that make up our brains,
humans deal most efficiently in simple concepts composed of few
elements interrelated in limited and well-defined forms: computers
are much better than humans at book-keeping masses of complex
data. Concordantly, and for a variety of reasons, the cultural
traditions of science now include the dictum that a simple explanation
is preferred over a complex one. Genetically and culturally, we inherit
parsimony as an essential sculptor of the scientific edifice:

A theory is the more impressive the greater the simplicity of its premises is,
the more different kinds of things it relates, and the more extended is its
area of applicability.

[A. Einstein (1949) 'Autobiographical Notes', *Albert Einstein:
Philosopher – Scientist* (P.A. Schlipp, ed.) Open Court, La Salle IL,
p. 33.]

For human reasons, parsimony should be an essential criterion in designing scientific explanations, and this is one half of Newton's first *Rule*: scientists should always strive for the simplest possible theories.

Parsimony is not a law of Nature

Newton's first Rule ascribes parsimony to all natural processes, but it is not at all clear that Nature always operates parsimoniously. Parsimony is an *extremum* principle: a parsimonious system is one in which the fewest possible causes produce the required effects. True, various *extremum* principles (such as the minimization of energy and the maximization of entropy) are often accurate approximations to the real world; even some *extremum* principles that are characteristically human in origin – certain criteria of elegance and simplicity – have led to remarkably useful models of natural phenomena. Nonetheless, the natural world is replete with examples of inefficient and overdetermined phenomena – phenomena in which one is hard pressed to see Newton's absolutely parsimonious Nature at work.

In some nervous systems, twice as many nerve cells are generated during embryonic development as actually survive to adulthood; the excess neurons simply die in their infancy and are never heard from again. Likewise, during its early development the human embryo grows a tail, but soon afterwards the entire tail degenerates through a massive wave of cell death. Or, consider the vertebrate eye, which is designed inside-out as compared to other eyes, such as the octopus eye.

(I have long been an admirer of the octopus. The cephalopods are very old, and they have slipped, protean, through many shapes. They are the wisest of the mollusks, and I have always felt it to be just as well for us that they never came ashore...

[L. Eiseley (1957) 'The snout', *The Immense Journey* Random House, NY, p. 47.])

In the octopus, light falls directly on the photoreceptors. In vertebrates, however, light must pass through many layers of cells and axons before it can reach the photoreceptor cells, which are then pointing in the wrong direction, away from the lens. Other dramatic examples of 'imperfection' can be found among the hormones and the neurotransmitters. Human engineers have created synthetic analogues of a wide range of biologically active molecules, where the analogues are hundreds or thousands of times more effective than Nature's originals. The *extremum* principle, if one exists at all, is well hidden in these situations – Nature does not operate as would a human engineer.

Nature's unparsimonious behaviour frequently confronts the biologist in the guise of unnecessary complexity: biological systems can contain more machinery than is necessary to make them work properly. The major role of DNA is to code for the production of proteins, yet 99 per cent of the DNA in human cells does not code for any protein. Cells in related animals can differ in their DNA contents by greater than one hundred-fold, and this suggests that if the excess, non-coding DNA plays any role at all it must be an exceedingly modest role at best. There is no simple rule for predicting how much DNA a cell will have; the actual DNA content of any particular cell is probably an accident of history, and the armchair biologist who attempts to construct a table of the DNA content in various cells sets forth on a futile exercise.

Another example of unnecessary complexity is the blood clotting cascade. When you cut your finger, blood proteins immediately begin to clump together, the wound is soon dammed up, and the cut stops bleeding within five to ten minutes. The initial injury sets off a waterfall of from eight to thirteen separate chemical reactions in two chain reactions, with each chemical transformation giving rise to the next chemical transformation in an orderly sequence. At least thirteen different proteins – coagulation factors – form the normal clotting cascades in humans, and if one of these factors is missing the person can have a bleeding disorder such as hemophilia. The complete blood clotting cascade is quite complex, and an armchair biologist would be hard pressed to predict its details from *a priori* considerations, from first principles, or from the requirements of the blood clotting system. One of the factors – Hageman Factor or Factor XII – even appears to be unnecessary: those people, who through genetic disorders, develop without any Factor XII do not have bleeding problems; and whales, dolphins and porpoises, which normally do not have any Factor XII, survive injuries quite normally.

The theoretician who tries to reason out such complex phenomena is like a parent who tries to deduce the route that his 6-year old has taken to school. It is always possible that the child walked straight there along the sidewalk, but it is equally likely that he detoured across the neighbor's backyard, stopped to inspect the trash pile at the side of the garage, crossed the street to throw a stick at a squirrel, talked with the mailman, kicked a stone into the bushes where he discovered a broken pencil, and finally marched into the school by the back door. Predicting children's behavior is a chancy business and, like abstractly explaining natural phenomena, it is an impossible task to carry off with perfection.

Rich and excess complexities permeate life. At the molecular level, there are the 'futile metabolic cycles' in cells, circular chemical reactions that go back and forth producing and unproducing the same molecules and depleting energy stores to no apparent purpose.

The two opposing reactions between glucose and glucose-6-phosphate (the hexokinase and glucose-6-phosphate reactions) would appear to cancel each other and result in the net dephosphorylation of ATP... Such a cycle is futile because it accomplishes nothing but the wasteful hydrolysis of ATP... [experimental analyses have shown that futile cycles] may go on simultaneously, sometimes at quite comparable rates.

[A.L. Lehninger (1975) *Biochemistry*, 2nd edn, Worth, NY, pp. 631–2.]

There is also the fact that many cells can use any one of a variety of fuels. The brain normally runs strictly on carbohydrates, burning the elemental sugar glucose for energy. During starvation, however, the brain takes advantage of other parallel metabolic pathways, switching over to burning ketones which are derived from fats. The chemical pathways of cells are not lean and geometric as if etched by a Piet Mondrian, rather they are varied, interdependent, and often overdetermined as if woven together by a Mark Tobey.

Complex and elusive intricacies characterize biological tissues. There is, for instance, the corpus callosum, one of the largest bundles of axons in the human brain. Although it extensively interconnects most areas of the cerebral hemispheres, its functioning is so subtle that for years no one understood exactly what it does. The five out of a thousand individuals born without a corpus callosum cannot normally be distinguished from those people with a corpus callosum; and most other animals do not have corpus callosums, living quite happily without them. It was only through an extremely specialized series of psychological experiments that Roger Sperry finally showed how the two halves of the human brain normally use the corpus callosum as their most intimate route of self communication.

From his desk, the theoretical biologist could not determine the role of the corpus callosum with certainty and he could not predict its appearance or its use in those animals (the placental mammals) that have acquired a corpus callosum during the last 200 million years. Faced with a plethora of possible designs, the armchair biologist cannot easily predict which ones will actually arise in the natural realm.

> What might have been is an abstraction
> Remaining a perpetual possibility
> Only in a world of speculation.
> What might have been and what has been
> Point to one end, which is always present.
> [T.S. Eliot 'Burnt Norton' from: *Four Quartets*.]

Who could have imagined that the human brain contains two separate minds, a right mind and a left mind, each localized in one of the major cerebral hemispheres? Normally, the two minds are in such close touch that they think alike, they trade thoughts instantaneously, they share the same sensations and emotions, and they act as one. All this intimacy flows through the corpus callosum, and the intercommunication is smooth and efficient; yet, at the same time, each separate brain is a powerful and complete mind. Amazingly, without a corpus callosum, the nervous system still functions as a smooth and efficient unit – one brain, to almost all outward appearance. Normally, two brains make each human, and two brains are a wonderful but unnecessary complexity.

Nature need not adhere to our standards of parsimony, and she need not follow our *exremum* principles. Nature does as she does, by accident, by wandering and by tinkering, and she has no teleology. We can only be secure in our science when we act as natural historians, conscientiously describing the natural realm retrospectively; we can rarely be confident armchair theoreticians; and we walk a precipitous course when we attempt *a priori* evaluations based on anthropocentric standards of parsimony.

Nature is not constrained to absolute parsimony, and the theoretician must be careful in using *extremum* principles. A scientist should not confuse parsimonious explanations with the checkered parsimonious and nonparsimonious behaviors of Nature.

Certain patterns must be completely templeted

The first principle of templeting reminds us that all patterns can be constructed from a completely explicit blueprint, even when such detailed specifications are not actually necessary, and the nonparsimonious behaviour of Nature demonstrates that such overdetermined explanations can be good explanations for certain natural phenomena. The second principle of templeting points out that a complete specification is at times an inescapable necessity.

A simple demonstration of this principle begins with $E = 2^N$, the number of equivalent templets, where N is the number of independent zeros in the matching matrix. (See the Appendix for another mathematical demonstration of the second principle.) When the raw materials are topologically naive, the matching matrix is completely filled with ones, $N = 0$, and $E = 1$ – there is only one appropriate templet. Can this templet T differ from the maximal templet, AM? We know that:

$$MM \wedge T = AM$$

and that the addition table for $\wedge$ includes the following relations:

$$1 \wedge 0 = 0$$
$$1 \wedge 1 = 1$$

Because of this table, T makes an exact copy of itself when MM is filled entirely with ones. Thus, for a completely nonrestrictive matching matrix:

$$MM \wedge T = T$$

and T must equal AM. When the matching matrix is completely nonrestrictive, there will be only one appropriate templet, which must be the maximal templet. In accord with the second principle of templeting, determinate pattern assembly with topologically naive raw materials always requires a completely explicit templet. When the 'universal assembly laws' are extremely permissive, extensive specific situational constraints become necessary.

DNA: the archetypic templet

The replication, the transcription, and the translation of DNA provide dramatic examples of the need for maximal templets in assembling certain complex natural patterns. DNA molecules are long strings of subunits, and these subunits (nucleotides) come in four different varieties: adenine, cytosine, guanine, and thymine. The four nucleotides are topologically naive – they can be strung together in any combination – and the almost limitless variety of DNA sequences is directly responsible for the almost limitless variety of organisms.

Encoded in their nucleotide sequences, DNA molecules contain much of the heritable information of organisms. There are three major steps in the expression of the encoded information, and each step is carried out by a particular set of enzymes. First, there is replication, in which a complementary DNA copy is made from a precursor DNA strand; the major enzymes at work in replication are DNA polymerases. Second, there is transcription, in which a complementary RNA copy is made from a precursor DNA strand; the major enzymes at work in transcription are RNA polymerases. Third, there is translation, in which a protein molecule is made from a precursor RNA strand. Proteins are assembled on tiny intracellular machines called ribosomes, under the guidance of a host of special enzymes.

All three decoding processes rely on maximal templets, and all three produce completely templeted patterns. In replication, for instance, the two complementary DNA molecules are unwound from their double helical coil and new complementary chains are constructed, nucleotide by nucleotide. Each adenine in the preexisting chain templets the addition of a thymine in the new chain, each thymine templets the addition of an adenine, each cytosine templets the addition of a guanine, and each guanine templets the addition of a cytosine. The new chain becomes an exact complement, a 'negative', of the templet chain and it becomes an exact copy, a 'positive', of the complement to the templet chain.

The 'universal laws' for nucleotide assembly allow sequences to be built in any order and natural DNA patterns are archetypic templeted patterns; templeting is also an important mechanism in the fabrication of many other biological patterns, from molecules to organelles to teeth to spots. The discovery that a particular natural pattern is assembled via a templeting mechanism can now be turned around and used to predict the interactive potentials of the precursor elements. Topologically naive precursors can only be determinately assembled into patterns via maximal templets, and therefore the templeted patterns include all patterns that are built from topologically naive precursors. In this way, when he discovers that a pattern is completely templeted, the biologist can immediately hypothesize that the constituent elements are topologically naive and can potentially form a great many different patterns. The lack of parsimony in many natural systems, however, dictates that this hypothesis must always be experimentally tested.

Other templets in the real world

Templets pervade the real world, and there are many examples of maximal templets – templets specifying the entire topology of a pattern. Large illuminated signs composed of hundreds of light bulbs forming designs or advising us to EAT AT JOE'S are completely templeted; the bulbs can potentially be arranged into any order, and the particular pattern of interest is built on a specific, completely prespecified templet of sockets. Hooked rugs are templeted; the strands of yarn can potentially be woven into any imaginable design, and the actual design to be stitched is prespecified by an explicit drawing on the background canvas. The letters of a crossword puzzle can, in theory, be arranged in almost any order, but each particular crossword puzzle is templeted by a set of blank squares and by the accompanying word clues. The notes of a piece of music can be juxtaposed in any possible order, but the detailed pattern of J.S. Bach's *Well-tempered Clavier* is completely prespecified in the templet of the musical score. Business forms comprise a set of organized, labelled blanks, such as: 'Name _____, Date_____,' etc. The words that fill these forms can potentially be arranged in any order on a blank page, but the particular form at hand templets the words into one particular pattern.

Templets arise in all disciplines that must make sense of complex systems. For example, ethology proposes that animal behaviors are fabricated of elemental behavioral units:

> Like the stones of a mosaic, the inherited and acquired elements of a young
> bird's behavior are pieced together to produce a perfect pattern.
>
> [K.Z. Lorenz (1952) *King Solomon's Ring. New Light On Animal
> Ways* (trans. M.K. Wilson) Thomas Y. Crowell, NY, p. 132.]

It appears that very young animals form neural templets as copies
of sensory experiences and that these 'sensory templates [then] guide
the development of certain kinds of motor behaviour' such as the
production of species-specific birdsongs [P. Marler (1976) 'Sensory
templates in species-specific behavior' In: J.C. Fentress (ed.)
Simpler Networks And Behaviour Sinauer, Sunderland MA, p. 328].
Ethologists propose that the elements of the final behavior can
potentially be arranged in many possible orders and that the mature
behavior is largely templeted by early experience. Similarly, metal-
lurgists explain much of the micro- and macrostructure of metals in
terms of templets. Alloys are formed by elemental molecular subunits
nucleating on a growing substrate, and the complex semicrystalline
patterns of metals result from the interplay of the intrinsic binding
preferences (self-assembly) of the molecular subunits and the detailed
surface topography (templet) of a particular substrate.

Q: the templeting index

How might we quantify the amount of templeting required
by various patterns? The appropriate measure should be maximal
when a maximal templet is required, it should be minimal when only
a minimal templet is required, and, to compare patterns of different
sizes, the measure should be normalized to the size of the pattern.
A simple measure that fulfils these requirements is Q, the templeting
index:

$$Q = S/S_{max}$$

with the limits:

$$\left(\sum_{row} AM \right)/n^2 \leqslant Q \leqslant 1$$

Templets are the explicit organizational instructions for pattern
assembly, and Q indicates what proportion of the maximal possible

templet is required to construct a particular pattern. If Q is much less than one, then the underlying assembly laws dictate the final configuration and only a minimal templet is needed; if Q equals one, then the underlying universal assembly laws are not restrictive, the final pattern is completely templeted, and a maximal templet is required.

The templeting index quantifies the extent to which the organization – the form – of a pattern must be explicitly prespecified during its construction, and it is an integral part of the explanation of any pattern. When the templeting index is minimal, the pattern can be formed from simple beginnings – i.e., intrinsic self-assembly laws. In contrast, when the templeting index is maximal, an explicit representation of the entire configuration is needed to build the pattern, and the pattern is a paragon of the second principle of templeting. Such patterns cannot be fabricated from simple beginnings; large or small, they must inherit their complete design.

> I found a dimpled spider, fat and white,
> On a white heal-all, holding up a moth
> Like a white piece of rigid satin cloth –
> Assorted characters of death and blight
> Mixed ready to begin the morning right,
> Like the ingredients of a witches' broth –
> A snow-drop spider, a flower like froth,
> And dead wings carried like a paper kite.
>
> What had the flower to do with being white,
> The wayside blue and innocent heal-all?
> What brought the kindred spider to that height,
> Then steered the white moth thither in the night?
> What but design of darkness to appall? –
> If design govern in a thing so small.
> [Robert Frost 'Design'.]

5

Self-assembly

The notion that mice can be generated spontaneously from bundles of old clothes is so delightfully whimsical that it is easy to see why men were loath to abandon it. One could accept such accidents in a topsy-turvy universe without trying to decide what transformation of buckles into bones and shoe buttons into eyes had taken place. One could take life as a kind of fantastic magic and not blink too obviously when it appeared, beady-eyed and bustling, under the laundry in the back room.

[L. Eiseley (1957) 'The secret of life' *The Immense Journey* Vantage, NY, p. 197.]

Self-assembly in general

To know a pattern is to know its roots – to understand a pattern, we must understand its history. Configurational explanations are historical explanations of pattern topology. Formally, a configurational explanation is a topologically naive automaton – a logical adder – that combines a set of raw materials and a templet to produce a final pattern. When the raw materials are topologically naive, the templet must contain significant explicit form information and the final pattern is extensively templeted. In contrast, when the raw materials are themselves topologically knowledgeable and allow the assembly of only one or a very few different configurations, the templet need not provide much additional form information and the final pattern is self-assembling.

A key and a lock, a hand and a glove, and a three-pronged plug and an electrical socket are all self-assembling units – the underlying 'universal assembly laws' are restrictive, and they can be fitted together in only one way. Carburetors, water faucets, lawn mower gas engines and wrist watches are self-assembling, each being composed of elements that can be fitted together into only one functional unit. A jigsaw puzzle cut in uniquely-shaped pieces can only be assembled into a single whole pattern. Various biological molecules, such as the collagens and many viruses, can, under physiological conditions, assemble into only one distinct configuration.

In self-assembling systems, the precursor elements embody sufficient information to uniquely specify a particular final pattern, and configurational explanations need not turn to detailed outside sources of topological information. Here, a topologically naive automaton can faithfully reproduce a particular configuration from a given set of raw materials using only a minimal amount of additional instructions, a minimal templet.

How big is a minimal templet? When the precursor elements are topologically knowledgeable, they can uniquely determine the final configuration and the matching matrix MM is identical to the adjacency matrix AM. If $MM = AM$, then the minimal necessary templet size is:

$$S_{min} = \sum_{row} AM$$

(S_{min} denotes the minimal necessary templet size for a minimal templet.) In such cases, the number E of equivalent templets is:

$$E = 2^{(S_{max} - S_{min})}$$

where S_{max} is the size of the adjacency matrix and S_{min} is the number of matrix slots actually filled by ones. Unless the final pattern actually has all of the possible interconnections between elements, minimal templets will always be members of a large family of equivalent templets. For a minimal templet, Q_{min}, the templeting index, is:

$$Q_{min} = S_{min}/S_{max}$$

and the patterns for which $Q \longrightarrow Q_{min}$ are the self-assembling patterns.

Self-assembly in the biological world

Organisms are patterns of matter that are at once complex and individual and recurrent, and the features that distinguish these complex and individual patterns from the other complex and individual patterns of the natural world are the two sequential processes – ontogeny and phylogeny – that produce organisms.

Ontogenies

Ontogeny is the developmental sequence of an organism. It is the history of a living entity from conception through birth to maturity and death, transforming an unspecialized embryonic form into a particular and idiomatic machine. Ontogenies come in all shapes and sizes. At one extreme, bacteria go through an ontogeny that is entirely internal; the transformations from a single parent cell to two daughter cells are all cascades of changes of molecules inside the cell. At the other extreme, multicellular organisms, such as worms, spiders and monkeys, begin as single cells – zygotes – and transform into unified collections of millions of cells through cascades of intracellular, cellular, and extracellular changes that establish whole cities of specialized cells. During the ontogeny of a multicellular organism, enclaves of cells are geographically segregated into organs and tissues connected by highways of nerves and vessels. The construction of this geography is continuously dynamic, and it proceeds inexorably in a particular sequence, the characteristic ontogeny of that organism.

Ontogenies, the building stages of organisms, are filled with self-assembling patterns. At the molecular level, there are the collagens, a major class of connective tissue (extracellular) molecules that are widely distributed throughout the animal kindom. All collagens are constructed on the same plan: three helical subunit chains are themselves wound together in a superhelix forming the basic rod-shaped collagen molecules, which are then longitudinally aligned as cross-linked fibrils. As with other polypeptides, collagen subunits are manufactured as linear strings by the intracellular ribosomes. Once fabricated, the inherent structure of these linear strings is entirely sufficient to fold them into the appropriate three-dimensional shape, to arrange them into fibrils, and to form all of the necessary cross-linkages inside and between fibrils. All of this spontaneous self-assembly can even take place outside of cells (in a suitably physiological environment). By the criterion of reproducible formation of a single distinct configuration, collagen fibers are certainly self-assembling systems.

The classic example of ontogenetic self-assembly is the tobacco mosaic virus. This virus is a tiny cylinder made of one molecule of RNA coated with 2130 identical protein subunits. The tobacco mosaic virus autonomously assembles inside cells, and identical viruses autonomously assemble in acellular physiological solutions. The tobacco mosaic virus is a single and specific complex structure – an intricate biological crystal – that reproducibly and autonomously coalesces through the interaction of a large number of precursor elements; it is an archetypic self-assembling system.

Phylogenies and simple origins

Generation after generation, ontogenies are reproduced as organisms beget like organisms, and this is phylogeny, the ancestral lineage of organisms. Ontogenies are the life histories of individual organisms, and a phylogeny is the repeated unfolding of ontogenies. Our ancestors are our phylogeny.

Biological time is different from physical time, and the phylogenetic clock ticks in generations. The physical time scale of phylogenies ranges through six orders of magnitude: ten human generations take two centuries, ten buttercup generations take a decade, ten fruit fly generations take six months, and ten bacterial generations can unfurl in three hours.

Biologically, a human phylogeny of ten generations is a very short time and it represents an almost unchanging set of transformations. In the course of two hundred years, humanness is preserved; those differences that do show up during a few generations are rather subtle, and each child is much more like his parents then he is different from them. A few generations of phylogeny is a biologically stable time interval.

Twenty million generations is quite another story – twenty million generations ago, humans were not human. At that time, our ancestors were $3\frac{1}{2}$–5 feet high ape-like creatures living in what is now eastern Africa. They were bands of primates, no taller than our children and certainly not as smart, living in forests and eating roots, tubers, and grubs. Although these animals looked much like apes, they differed from their neighboring apes in a dramatic and significant way: twenty million generations ago, the members of the human phylogeny – *Australopithecus afarensis* – walked upright.

In phylogeny, notable changes happen on a time scale of millions of generations, and ten million generations after *Australopithecus afarensis* the new members of the human phylogeny – *Homo habilis* –

had acquired the large brains that are characteristic of today's humans. In the course of millions of generations, the differences between members of a phylogeny can become so marked that the original organism evolves into a new organism, and over the course of ten million years *Australopithecus* evolved into *Homo*.

Phylogenies slowly but resolutely evolve, and evolution is never ending:

> It gives one a feeling of confidence to see nature still busy with experiments, still dynamic, and not through nor statisfied because a Devonian fish managed to end as a two-legged character with a straw hat. There are other things brewing and growing in the oceanic vat. It pays to know this. It pays to know there is just as much future as there is past. The only thing that doesn't pay is to be sure of man's own part in it.
>
> There are things down there still coming ashore. Never make the mistake of thinking life is now adjusted for eternity. It gets into your head – the certainty, I mean – the human certainty, and then you miss it all: the things on the tide flats and what they mean, and why, as my wife says, 'they ought to be watched.'
>
> [L. Eiseley (1957) 'The Snout' *The Immense Journey* Vantage, NY, pp. 47–8.]

Evolution means change, and typical phylogenetic questions are about changes in the lineage of organisms: 'Why did elephants acquire trunks?' 'How did dinosaurs become extinct?' 'When did mammals develop placentas?' 'From whence came mitochondria?' Complete answers to such 'change' questions have three parts.

First, all evolutionary changes must be heritable changes; therefore, one part of the answer is a description of the genetic change (mutation, rearrangement, duplication or deletion of the genome) that occurred. Second, the change had a particular cause – a mutagen (such as excess radiation), an inherent tendency for the genome to rearrange itself (a 'jumping gene'), or human intervention (such as a recombinant DNA experiment). What was that cause? Third, the genetic change was long-lasting; it became sufficiently entrenched to be part of a recognizable lineage of organisms. The classic Darwinian explanation for entrenchment is that the initial genetic change produced a selective advantage for the organism. Another explanation is that genetic stability is the norm and that, once a change occurs, it tends to be preserved within the genome by natural, intrinsic buffer mechanisms. In any case, the third part of a complete answer to questions about particular phylogenetic changes is an explication of the persistence of the underlying genetic change.

[Evolutionary biologists usually attend to changes, but the stability of lineages is equally striking. Phylogenetic changes tend to be fairly subtle variations on basic stable themes of morphology and of biochemistry. The elephant trunk is a geometrically larger but topologically identical nose to that of other ungulates. The new tetrameric (four-part) hemoglobin molecule of mammals is an assemblage of variants of the much older monomeric (one-part) myoglobin molecule of primitive fish. It is the constancies and the similarities preserved during long phylogenies that are responsible for the overall stately pace of evolution.]

One of the most difficult evolutionary questions is the elemental problem of how life originated. Perhaps, somewhere in the primal soup, complex macromolecules slowly progressed from inanimate, to semi-animate, to animate. But, what would semi-animate molecules look like, how would they behave, and could they have persisted for any significant period of time? Or, was there a rare and chance event that suddenly coagulated an inanimate precursor into the first squiggling organism, making a direct jump from the inanimate to the animate – from the simple to the complex – investing some group of dead precursors with just the appropriate complex order to make a living being?

At base, these questions ask: How much of the origin of life was templeting and how much was self-assembly? and the implicit difficulty in answering stems from the templeting end of the spectrum. Self-assembly does not fully explain the organisms that we know; contemporary organisms are quite complex, they have a special and an intricate organization that would not occur spontaneously by chance. The 'universal laws' governing the assembly of biological materials are insufficient to explain our companion organisms: one cannot stir together the appropriate raw materials and self-assemble a mouse. Complex organisms need further situational constraints, and, specifically, they must come from preexisting organisms. This means that organisms – at least contemporary organisms – must be largely templeted.

Today's organisms are fabricated from preexisting templets – the templets of the genome and the remainder of the ovum – and these templets are, in turn, derived from other, parent organisms. The astronomical time scale of evolution, however, adds a dilemma to this chain-of-templets explanation: the evolutionary biologist presumes that once upon a time organisms appeared when there were no preexisting organisms. But, if all organisms must be templeted, then what were the primordial inanimate templets, and whence came those templets?

The dilemma of the evolutionary beginnings of life originates in a philosophical issue central to scientific explanations of any complex form. Templeting can explain how a complex pattern arises, but templeted patterns are patterns with complex origins and the explanation of complexity on the basis of complexity is not satisfying. Self-assembling patterns, on the other hand, can be reduced to inherent 'universal laws' and simpler predecessors, and reduction to simpler origins is the traditional scientific paradigm of explanation. Scientists prefer self-assembly rather than templeting as an ultimate explanation, and, for this reason, the self-assembling aspects of natural phenomena have received a disproportionate amount of scientific attention.

Why this preference? Why prize the simple over the complex? It is likely that the human brain dictates our need for simplicity. It is easier for humans to remember an explanation that has few parts than to remember an explanation with many parts, and it is easier for us to manipulate an abstraction with few and uniform inter-linkages than to wrestle with an unwieldy abstraction tied together like an exploded Christmas tree.

Given that simple explanations are better then complex explanations for humans, why entertain complex explanations at all? There are two important reasons. First, to a computer, complex explanations can be as useful as simple explanations – computers can deal with large and convoluted abstract structures as comfortably as with small and smooth abstractions; computers are more limited by the precision with which an abstraction is formulated than by its size or its heterogeneity. In an era when machine representations of phenomena are increasingly interwoven into the fabric of synthetic science, complex explanations, as well as simple explanations, can be a significant part of scientific theories.

Second, complex explanations are inescapable in the natural world.

Certain phenomena cannot be directly reduced to smaller, simpler, or less intricate origins. The intrinsic 'universal assembly laws' are sometimes insufficient; some real world patterns, like natural DNA molecules, must be extensively templed; and the second principle of templing reminds us that complex origins are a fact of life. Still and all, perhaps a scientist should not rest on a scientific explanation that reduces to a complex templet. Perhaps he should doggedly pursue his explanation backward, hoping to finally discover distant predecessors that truly are simpler in an ultimate explanation that is self-assembling. This avenue is always open to the scientist, but the eventual and distant reduction of an extensively templed pattern to simpler origins does not undo the fact that the actual pattern was directly fabricated from an intricate and complex predecessor. The first level configurational explanation provides the fundamental and relevant insight into the immediate pattern-assembly process, and for this reason the first level configurational explanation – complex though it may be – cannot be disregarded.

I suspect that the urge to continue reducing phenomena toward simpler origins also takes courage from a secret belief that there is always an ultimate reduction that will most truly explain things. This may be. I am, however, of the Eiseleyian persuasion, who wrote:

I do not think, if someone finally twists the key successfully in the tiniest and most humble house of life, that many of these questions will be answered, or that the dark forces which create lights in the deep sea and living batteries in the waters of tropical swamps, or the dread cycles of parasites, or the most noble workings of the human brain, will be much if at all revealed. Rather, I would say that if 'dead' matter has reared up this curious landscape of fiddling crickets, song sparrows, and wondering men, it must be plain even to the most devoted materialist that the matter of which he speaks contains amazing, if not dreadful powers, and may not impossibly be, as Hardy suggested, 'but one mask of many worn by the Great Face behind.'

[L. Eiseley (1957) 'The secret of life' *The Immense Journey* Vantage, NY, p. 210.]

6

Rules for configurational explanations

Will Strunk loved the clear, the brief, the bold, and his book is clear, brief, bold. ... He scorned the vague, the tame, the colorless, the irresolute. He felt that it was worse to be irresolute than to be wrong. I remember a day in class when he leaned far forward, in his characteristic pose – the pose of a man about to impart a secret – and croaked, 'If you don't know how to pronounce a word, say it loud! If you don't know how to pronounce a word, say it loud!' This comical piece of advice struck me as sound at the time, and I still respect it. Why compound ignorance with inaudibility? Why run and hide?

[E.B. White (1979) *The Elements Of Style* by William Strunk Jr Macmillan, NY, p. xvi.]

The power of configurational explanations lies in their explicit separation of the precursor elements and the templets. To use them to best advantage, I suggest the following rules (which presume some consistent philosophical assumptions about inference, such as the set of five postulates proposed by Russell [R.E. Egner & L.E. Denonn (1961) 'Non-demonstrative inference' *The Basic Writings Of Bertrand Russell* Simon & Schuster, NY, pp. 655–66]):

1. *Define the topology of the pattern*

A. *Choose a substantive pattern*

For a configurational explanation to be predictive, the pattern that it explains should be coherent and relatively permanent. When a pattern is fragmented or ephemeral, formulating configurational explanations can be an empty exercise. Reproducibility is a good indication that a pattern is substantial: If the pattern can be created on demand, is its reproduction repeatedly accurate? If the pattern is collected by observation rather than generated by experiment, does it recur with high fidelity?

B. *Partition the pattern into discrete elements*

Descartes' second rule of reasoning was to divide each phenomenon into many parts – follow Descartes and carve your pattern into discrete units, apportioning boundaries and edges. The partitioning should be done with an eye to the fact that the elements serve double duty. First, they are the natural units of the final pattern. In addition, these elements are the precursors from which the pattern is assembled: they must be able to exist independently, and they must exist or they must be created before the final pattern is fully assembled.

C. *Define a binary relation between the elements*

Find or invent a relation, m, that any two elements in the pattern either definitely share or definitely do not share. As Eddington wrote:

> We take as building material *relations* and *relata*. The relations unite the relata; the relata are the meeting points of the relations. ... I do not think that a more general starting-point of structure could be conceived.
>
> [A.S. Eddington (1929) *The Nature Of The Physical World* Macmillan, NY, p. 230.]

D. *Abstract the pattern*

Now, model the pattern entirely within the abstract realm. For a configurational explanation, abstract the pattern as a graph of its elements interlinked by the defined binary relation, m, and then summarize the graph as an adjacency matrix, AM, that lists all of the constituent elements and that classifies their interconnections as ones or zeros. The adjacency matrix is a succinct definition of the topology of the pattern.

2. *Describe the innate topological potentials of the raw materials*

Next, compose an abstraction of the precursor elements and summarize this precursor abstraction as a binary matrix – a matching matrix, MM. The matching matrix classifies all of the potential interconnections between precursor elements in terms of the chosen binary relation, m. The precursor elements must have the potential to be interconnected into the final pattern; thus, the matching matrix always contains the adjacency matrix. To define the other potential interconnections (those not found in the final pattern):

    (a) Observe a variety of naturally-occurring patterns in the real world to determine whether other interconnections ever arise.

    (b) Experiment to determine whether other interconnections are ever possible.

The matching matrix is a standard and concise summary of the topological information inherent in the raw materials, and it embodies the 'universal assembly laws' you allow for your system.

3. Characterize the minimal necessary templets

The first principle of templeting points out that all patterns can be constructed from a maximal templet, an explicit image of the pattern itself. Is that actually necessary for your pattern? Must the complete configurational information be explicitly prespecified? To find out, characterize the equivalence class of appropriate templets by determining which templet matrices, T, fulfill the equation:

$$MM \wedge T = AM$$

Compute the minimal necessary (informational) size S, the number E of equivalent templets, and the templeting index Q. The templeting index tells you what proportion of the maximal templet is minimally necessary to assemble your particular pattern, and it is a measure of the extent to which your pattern is determined by *ad hoc* situational constraints.

4. Find the real templet

Abstract templets are predictions about the real world. If your configurational explanation is a good model, then the predicted templets should be good scientific abstractions of particular real world items.

Look for a real templet, things that interact with the raw materials to organize the form of your pattern. Start with the prediction that only the minimal templet is actually used. The minimal templet is both necessary and sufficient, but, in a world without parsimony, necessity and sufficiency cannot form the complete basis of secure prognostication. True cause may go beyond necessity and sufficiency. In the end, to determine the real templet, you must go out into the world and find the real templet.

If your abstract templets cannot be well matched to real items, then your configurational explanation is not a good approximation to the real world. In this case, begin again with step 1 and modify the choices that you made along the way. The art of making these choices is 'plausible reasoning' or, more simply, educated guessing:

To a philosopher with a somewhat open mind all intelligent acquisition of knowledge should appear sometimes as a guessing game, I think. In science as in everyday life, when faced by a new situation, we start out with some guess. Our first guess may fall wide of the mark, but we try it and according to the degree of success, we modify it more or less. Eventually, after several trials and several modifications, pushed by observations and led by analogy, we may arrive at a more satisfactory guess. The layman does not find it surprising that the naturalist works in this way. ... It may appear a little more surprising to the layman that the mathematician is also guessing. The result of the mathematician's creative work is demonstrative reasoning, a proof, but the proof is discovered by plausible reasoning, by guessing.

[G. Polya (1968) *Mathematics And Plausible Reasoning. Vol. II. Patterns Of Plausible Inference* Princeton Univ. Press, p. 158.]

Remember, creating good scientific explanations is an algorithm of educated iterative guesswork. Science

does not grow through a monotonous increase of the number of indubitably established theorems but through the incessant improvement of guesses by speculation and criticism, by the logic of proofs and refutations.

[I. Lakatos (1976) *Proofs And Refutations. The Logic Of Mathematical Discovery* (ed. J. Worrall & E. Zahar) Cambridge Univ. Press, p. 5.]

7

Simple, complex and random

One might say, from this more objective standpoint, we were two organisms. Two of those places where the universe makes a knot in itself, short-lived, complex structures of proteins that have to complicate themselves more and more in order to survive, until everything breaks and turns simple once again, the knot dissolved, the riddle gone.

[L. Gustafsson 'Elegy for a dead labrador', (trans: Y.L. Sandstroem) New Yorker, August 24, 1981, p. 34.]

Simple and complex are two ends of a spectrum. Simple things have few parts, the parts are organized in a homogeneous fashion, and the whole can be fairly easily grasped in one fell swoop. Complex things have many parts, the parts are organized heterogeneously, and it takes a concentrated effort to comprehend the whole. A musical scale is simple, a Bach fugue is complex. A color chart is simple, a Jackson Pollock painting is complex. The alphabet is simple, the Bible is complex.

At times, the simplicity or complexity of an item can be deceptive. We sit down to a steaming meal of clams, lobster tails, crab legs, chicken, and vegetables. 'Delicious, but obviously complicated to make,' we comment to our host. 'Not at all complicated,' says he. 'Your simply steam all of the ingredients in a large pot for an hour. The meal cooks itself.' On the other hand, the first course was quenelle de brochette – egg-shaped fish mousses, looking like tiny homogeneous puddings. 'And the quenelles,' we ask, 'did you simply mix and cook those too?' Our host fixes us with an irritated stare; 'The quenelles took my full efforts for the entire afternoon.'

The effort and the intricacy of its fabrication can belie the apparent simplicity or complexity of a final creation. One might say: 'Simple is as makes itself, but complex needs two hands' – a good indicator of the most essential complexity of something is the difficulty of its construction.

The complexity of patterns

Physical size

As normally used, 'complex' has more than one meaning. On the one hand, it sometimes denotes multipartite: of two otherwise equivalent patterns, the more extensive can be called the more complex. Given the patterns:

1-*a*-1-*a*   1-*a*-1-*a*-1-*a*-1-*a*-1-*a*-1-*a*

the longer sequence can be called the more complex of the two.

Heterogeneity of content

'Complexity' can also mean the variety of constituent elements. Consider the three patterns:

1-1-1-1 $<$ 1-*a*-1-*a* $<$ 1-*a*-*-#

ranked according to their heterogeneity or complexity of content. The first pattern is the simplest because it contains the fewest different elements, and it can be described by a short statement, such as: 'four ones'. The second pattern is more complex; it contains more different elements, and its short descriptor is a bit longer: 'two ones and two *a*'s, alternating'. The third pattern is the most complex; it contains the most different elements, and its short descriptor requires a complete listing of the pattern: '*one* followed by *a* followed by* followed by #'.

Heterogeneity of configuration

A pattern can be naturally divided into its content and its configuration, and the 'complexity' of a pattern can equally well describe the heterogeneity of its configuration. Of the three patterns:

1-1-1-1 $<$ 1-1-1-1 $<$ 1-1-1-1

the first can be called the least complex because it has only one interconnection type – each element has two interconnections. The second pattern is more complex: it has two interconnection types – two elements have two interconnections and two elements have three interconnections. The third pattern is the most complex, having three interconnection types – one element has one interconnection, two elements have two interconnections, and one element has three interconnections.

These three types of complexity – physical size, heterogeneity of content, and heterogeneity of configuration – are all properties of the final pattern by itself. Each is the complexity of an isolated pattern, independent of its neighbors, its future, and its past.

Complexity of fabrication

No real world patterns stand in frozen isolation, icy absolute sculptures waiting outside of time. Real patterns are dynamic; they have ancestors, and they give rise to progeny. Essential descriptions of a pattern must reflect its life, and the *essential* complexity of a pattern should be an active quality.

To characterize a pattern dynamically, within its historical context, we can assess the difficulty of its construction – its complexity of fabrication. How much effort went into building it? One straightforward measure of this effort is the amount of information needed to assemble the pattern. Technically, the information content I of something is logarithmically related to the probability p of its occurrence:

$$I = -\log_2 p$$

The probability that a logical adder (the automaton representation of a configurational explanation) will produce a particular output pattern is:

$$p = 2^{-S}$$

where S is the minimal necessary templet size for that particular pattern. Thus, the information needed to assemble the pattern is simply equal to S.

S, the information content of a pattern, also makes a very useful measure of the complexity of fabrication. The essential complexity of a pattern – its complexity of fabrication – is the size of the minimal templet needed for its construction. Simple patterns need only small prepatterns; complex patterns require extensive and detailed blueprints.

Formal definitions of complexity
Traditionally, complexity and its obverse, simplicity, have posed definitional problems.

The problem of simplicity is of central importance for the epistemology of the natural sciences. Since the concept of simplicity apears to be inaccessible to objective formulation, it has been attempted to reduce it to that of probability...

> [H. Weyl (1949) *Philosopy Of Mathematics And Natural Science* Princeton Univ. Press, Princeton, pp. 155–6.]

Formal definitions of complexity are recent inventions, inspired by investigations into the theoretical underpinnings of the computer. In most technical settings, 'complexity' has come to denote a minimal necessary size or minimal necessary resources measure. Such definitions were foreshadowed by von Neumann who began by pointing out (in 1948):

We are all inclined to suspect in a vague way the existence of a concept of 'complication'. This concept and its putative properties have never been clearly formulated.

> [A.H. Taub, ed. (1963) *John von Neumann. Collected Works. Vol. V. Design of Computers, Theory of Automata and Numerical Analysis,* 'The general and logical theory of automata.' Macmillan Co., NY, p. 312]

By 'complication' von Neumann appears to have meant the heterogeneity of constituents and of internal interactions – in other words, complexity of content and complexity of configuration. He then went on to formally discuss automata that can produce other automata, and concluded:

'complication' on its lower levels is probably degenerative, that is, automata will only be able to produce less complicated ones. There is, however, a certain minimum level [of complexity] where this degenerative characteristic ceases to be universal. At this point automata which can reproduce themselves, or even construct higher entities, become possible. This fact, that complication, as well as organization, below a certain minimum level is degenerative, and beyond that level can become self-supporting and even increasing, will clearly play an important role in any future theory of the subject.

> [*op. cit.*, p. 318.]

In other words, von Neumann implied that the complexity of an automaton might be usefully related to the complexity of another automaton that constructed it.

The next major conceptual step was taken in the mid-1960s when A.N. Kolmogorov [(1965) 'Three approaches to the quantitative definition of information' *Problems of Information Transmission* **1**: 1–17] and ˙G.J. Chaitin [(1966) 'On the length of programs for computing finite binary sequences' *Journal of the Association for Computing Machinery* **13**: 547–69] each proposed that the essential complexity of a pattern is the size of the minimal precursor pattern – the minimal templet – necessary for its construction. Specifically, they rigorously defined the complexity of a binary string *A* as the size of the minimal binary string *B* that is the necessary input for a universal machine that outputs *A*. Next, they pointed out that a string can always be completely specified by a templet that is simply the string itself. Finally, they proposed that when the templet string *B* is approximately as long as *A*, then *A* should be considered to be maximally complex. By 'complexity', Kolmogorov and Chaitin mean 'complexity of fabrication'.

The importance of the Kolmogorov–Chaitin definition lies not only in its rigorous description of the complexity of fabrication, it also stems from the remarkable interrelationships that it suggests between all types of complexity, between the complexity of fabrication and the various heterogeneities of the final pattern. P. Martin-Lof [(1966) 'The definition of random sequences' *Information and Control* **9**: 602–19] has shown that those binary strings fulfilling the Kolmogorov–Chaitin requirements for maximal complexity also fulfill all definitions of random binary strings and that maximally complex binary strings possess as large a variety of different subsequences as would be expected in extremely heterogeneously ordered – i.e., randomly ordered – binary strings. The only difference between true random strings and other maximally complex strings is that random strings are generated by an unpredictable, stochastic process.

Kolmogorov's and Chaitin's idea that the essential complexity of a pattern is the size of its minimal necessary templet seems also to have motivated a variety of other characterizations of complexity. One notable example is Jeffreys' measure of the complexity of a differential equation [H. Jeffreys (1937) *Scientific Inference* Cambridge Univ. Press, ch. IV, pp. 36–51]. Jeffreys reduced all finite and practical (i.e., useful in a physical setting) differential equations to a universal form of finite order and degree and with integer constants. In this way, a scientist could (theoretically) assemble almost any differential equation by appropriately arranging a particular set of integer constants in the blanks in the standard form. The integer constants can potentially be inserted in any order: the precursor elements (the integers) are topologically naive, the final pattern of integer constants must be completely templeted, and the minimal necessary templet size is equivalent to the size of the final equation. Jeffreys proposed that the physical setting (the 'physical law') motivating the use of any particular differential equation determined the size or dimension of the equation. He then pointed out that if the underlying physical law required an equation with a standard form that is quite large – for which there are a great many blanks for numerical constants in the differential equation – then the complexity of that law should be considered to be quite high. In other words, Jeffreys suggested that when the minimal necessary templet is large, then the essential complexity is also large.

A similar perspective of complexity has been advanced by Popper [K.R. Popper (1959) *The Logic Of Scientific Discovery* Basic Books, NY, pp. 136–45]. Popper suggests that simple systems are easy to falsify while complex systems are difficult to falsify. From falsifiable systems, statements can be generated to distinguish among all of the conceivable system configurations – internal (or, as Popper calls them, 'universal') rules limit the possible states of the system and the system is self-assembling and relatively simple. In contrast:

Statements which do not satisfy the condition of falsifiability fail to differentiate between any two statements within the totality of all possible empirical statements.
[Popper, *op. cit.*, p. 92.]

Unfalsifiable, or very complex, systems generate a host of statements that cannot be distinguished by universal internal laws but which must be determined in each particular case by situational, extraneous, externally-imposed, and *ad hoc* constraints. Unfalsifiable, very complex systems are completely templeted.

Maximally complex patterns

may came home with a smooth round stone
as small as a world and as large as alone.
For whatever we lose (like a you or a me)
it's always ourselves we find in the sea.

(e.e. cummings 'maggie and milly and molly and may'.)

In the real world, seemingly complex patterns can be fabricated quite reproducibly – i.e., determinately. For instance, computers can repeatedly print the first 100000 digits of the number *pi* with high accuracy, and cells can faithfully replicate DNA sequences that are millions of subunits in length.

The instructions for a computer that prints a long sequence of *pi* can be significantly shorter than the number of digits actually printed, but the instructions for building a particular DNA molecule are always as long as the molecule to be built. In fact, it is an inherent property of certain very complex patterns (like many natural DNA sequences) that they can only be built from a templet that is at least as large as the pattern to be constructed. This is the second principle of templeting at work: certain patterns require maximal templets. From the Kolmogorov–Chaitin definition, these patterns can be considered to be maximally complex.

Maximally complex patterns are patterns necessarily built from maximal templets, and some real world patterns are inescapably maximally complex. An elemental law of pattern formation is that certain patterns cannot be directly reduced to smaller components: these patterns – the truly complex patterns – can only be explained in terms of predecessors (maximal templets) of the same size.

A maximal templet is equivalent to the final pattern. Both can have an identical content and configuration, and the maximal templet is the most explicit possible representation of the topology of the final pattern. Maximal templets are the raw and unabridged reflections of the forms and the order of complex patterns; they are the 'ourselves' that complex beings 'find in the sea'.

Random patterns

Some of the most complex patterns in the natural world are random patterns – the patterns of the waves, ripples, and sea-spray breaking along a jagged cliff on a windy day; the patterns of the clouds in late November; the patterns of the noon-day sounds in the center of a city. Random patterns are complex natural patterns produced stochastically. How might stochastic patterns be understood in terms of configurational explanations?

In a configurational explanation, a set of precursors combines with a templet to produce a particular pattern. When the real world pattern-assembly processes are largely determinate, the abstract templets of the configurational explanation correspond to coherent stable things in the natural world – the real world templets have a unity and a permanence.

On the other hand, when the pattern-assembly processes are largely stochastic, then the abstract templets represent ephemeral or unstable things in the real world. In stochastic processes, the real templets are recreated at each usage; they are transient, and they have no permanent coherence. The transience of the causal templets is the basis of modern physics' stochastic explanations of matter.

I give the name of 'causal line' to a series of events having the property that from any one of them something can be inferred as to neighboring events in the series. It is the fact that such causal lines exist which has made the concept of 'things' useful to common sense, and the conception of matter useful to physics. It is the fact that such causal lines are approximate, impermanent, and not useful which has caused modern physics to regard the conception of 'matter' as unsatisfactory.

[B. Russell (1961) 'Non-demonstrative inference' In: R.E. Egner & L.E. Denonn (eds.) *The Basic Writings Of Bertrand Russell* Simon & Schuster, NY, p. 652.]

Random patterns are produced by stochastic processes, and random patterns can be captured in configurational explanations containing ephemeral templets. (As a practical rule, to determine whether a given pattern-assembly system is determinate or stochastic, you can study the permanence of the real templets. And, when it is not easy to examine the real world templets, you can follow the fidelity of recurrence of individual final patterns – these will mirror the stability of their templets.)

Random patterns are very special, however, and all patterns produced by stochastic processes are not random. Randomness has two aspects: First, there is unpredictability; to say that coin-flipping produces random outcomes means that the final pattern of heads and tails is unpredictable. Second, there is heterogeneity; the distribution of heads and tails is also quite heterogeneous in any long sequence of coin-flips.

That these two aspects of randomness are actually separate characteristics of pattern-assembly processes is well demonstrated by the pseudorandom number generators. A pseudorandom number generator is a short algorithm that can output a long heterogeneous sequence of integers. [For details, see G.A. Mihram (1972) *Simulation. Statistical Foundations And Methodology* Academic, NY, pp. 44–56.] For instance, the algorithm:

$$r_{i+1} = (ar_i) \bmod b$$

can, with suitable constants a, b, and r_0, automatically generate long strings of random-looking numbers, $r_1, r_2, \ldots$ Setting $a = 203$, $b = 10^5$, and $r_0 = 8$, we get:

0225024173547526295516534 ...

By most tests – usually counts of the number of different sub-sequences and also counts of the number of times that each of these subsequences appears – pseudorandom numbers are extremely heterogeneous. Thus, pseudorandom numbers embody one of the two essential characteristics of random patterns: each long pseudorandom number is an extremely heterogeneous sequence of digits.

On the other hand, pseudorandom numbers do not have the second essential characteristic of random patterns: each pseudorandom number can be, in a very fundamental sense, completely predictable. By fixing the few input values for the algorithm, the same long pseudorandom number will always be output. Setting $a = 203$, $b = 10^5$, and $r_0 = 8$ in the above algorithm always produces the identical output sequence:

0225024173547526295516534...

Although pseudorandom numbers are extremely heterogeneous, they are nonetheless predictable because pseudorandom number generators are determinate machines not stochastic processes.

In terms of configurational explanations, the two aspects of randomness – heterogeneity and unpredictability – appear as complexity and stochasticism. A random pattern is one that is very complex – the underlying 'universal assembly laws' do not help us in understanding the pattern, the raw materials are topologically naive and a maximal templet is needed. In addition, truly random patterns are only produced by configurational explanations that represent stochastic pattern-assembly processes, configurational explanations with ephemeral stochastic templets. (For full details, see the Appendix.)

Life

Pick up an introductory high school or college textbook on biology, and you are likely to find the following definition of life:

Life is matter that exhibits the properties of metabolism, growth, movement, irritability, and reproduction.

Or stated more succinctly:

Life is the activities of protoplasm.

Such definitions are a beginning. They catalogue some of the important characteristics of most organisms – elephants, sponges, eucalyptus trees, and bacteria – and they describe common, special properties of those entities clearly acknowledged to lie within the biological realm. Nevertheless, these definitions are not as helpful in the more difficult cases such as viruses and self-reproducing automata.

Viruses are large, very specific, highly reproducible (recurrent) macromolecular complexes that parasitize cells. On their own, they neither metabolize, grow, move, respond to stimuli, nor reproduce. In conjunction with cells, however, they perform all of these functions, and the combination of virus–cell is certainly alive. Moreover, the virus is absolutely necessary for certain processes; the cell alone functions quite differently from the virus–cell. Viruses are near the border between life and non-life, but on which side of the fence are they?

Viruses are not self-sufficient, and perhaps the definition of life hinges on this feature. Consider, then, the (still theoretical) case of self-sufficient, self-reproducing automata. These machines, originally described by von Neumann, can extract the necessary materials from a nonliving environment and can construct exact copies of themselves. The copies then make more duplicates, and so on.

Self-sufficient, self-reproducing automata hold a special place in the science fiction literature; for example:

We have the planet Mars, a large piece of real estate, completely lacking in economic value because it lacks two essential things, liquid water and warmth. Circling around the planet Saturn is the satellite Enceladus.... composed of dirty ice and snow, with dirt of a suitable chemical composition to serve as construction material for self-reproducing automata. [We begin] with a rocket... launched from the Earth and quietly proceeding on its way to Enceladus. The payload contains an automaton capable of reproducing itself out of the materials available on Enceladus, using as energy source the feeble light of the far-distant sun. The automaton is programmed to produce progeny that are miniature solar sailboats.... The sailboats are launched into space... [carrying] a small block of ice from Enceladus. The sole purpose of the sailboats is to deliver their cargo of ice safely to Mars. ... A few years later, the nighttime sky of Mars begins to glow bright with an incessant sparkle of small meteors [i.e., millions of the solar sailboats].... Day and night the sky is warm. Soft breezes blow over the land, and slowly warmth penetrates into the frozen ground. A little later, it rains on Mars for the first time in a billion years. It does not take long for oceans to begin to grow. There is enough ice on Enceladus to keep the Martian climate warm for ten thousand years and to make the Martian deserts bloom.

[F. Dyson (1979) *Disturbing The Universe* Harper & Row, NY, pp. 199–200.]

Self-sufficient, self-reproducing automata are certainly tantalizing entities, but are they alive? They have most of the essential features of a typical organism: they reproduce, they grow, and they metabolize (they actively restructure environmental matter). Furthermore, to work properly, they must undoubtedly move. Nonetheless, self-sufficient, self-reproducing automata are machines, and can a machine really be alive?

The goal of definition is to capture the greatest number of attributes in the fewest possible statements, to build the simplest and the most encompassing scientific abstractions, and to do this by extracting the fundamentally distinguishing essence of the things to be defined. Problematic cases such as viruses and self-reproducing automata have led biologists toward revised and simplified definitions of life. Rather than describing the territory within which life resides, recent definitions directly address the borderland between life and non-life. These newer, simpler, and deeper definitions all include two characteristics: life exhibits a characteristic complex organization, and life

Life 83

exhibits a characteristic type of high-fidelity reproduction. For example, A.L. Lehninger wrote:

The most extraordinary attribute of living organisms is their capacity for precise self-replication, a property that can be regarded as the very quintessence of the living state. In contrast, collections of inanimate matter show no apparent ability to reproduce themselves in forms identical in mass, shape, and internal structure through 'generation' after 'generation'.

[A.L. Lehninger (1975) *Biochemistry*, 2nd edn Worth, NY, p. 4]

R.D. Hotchkiss [R.W. Gerard (ed.) (1958) 'Concepts of biology' *Behavioral Sci.* 3: 129] has concisely summarized these notions in his wonderful epigram: 'Life is the repetitive production of ordered heterogeneity.'

These recent and trenchant definitions suggest that the unique characteristics of organisms are *pattern* characteristics. The first of these fundamental pattern characteristics is complexity. Cells and organisms are quite complex by all pattern criteria. They are built of heterogeneous elements arranged in heterogeneous configurations, and they do not self-assemble. One cannot stir together the parts of a cell or of an organism and spontaneously assemble a neuron or a walrus: to create a cell or an organism one needs a preexisting cell or a preexisting organism, with its attendant complex templets. A fundamental characteristic of the biological realm is that organisms are complex patterns, and, for its creation, life requires extensive, and essentially maximal, templets.

The second essential and distinguishing feature is that the very complex patterns of life are recurrent and highly reproducible. As Schrodinger has pointed out in his essay *What Is Life? The Physical Aspect Of The Living Cell* (Macmillan, NY, 1945), a determinate or 'clockwork' mechanism characterizes life; the particular complex pattern that is a muscle cell or a spider is faithfully reconstructed time after time. For example, the frequency of errors in assembling long DNA molecules is less than one mistake per 3×10^4 subunits. From the perspective of configurational explanations of complex patterns, this fidelity of pattern-assembly puts biological processes on the highly determinate end of the spectrum. Living patterns are formed by pattern-assembly processes that are effectively determinate.

The two features – maximally complex patterns and determinate pattern-assembly processes – summarize the essence of the biological realm. In terms of configurational explanations, one might say:

Life is characterized by maximally-complex determinate patterns, patterns requiring maximal determinate templets for their assembly

and, a biologist is a scientist who luxuriates in recurrent complexity. Random patterns also require maximal templets, but random patterns are produced via stochastic pattern-assembly processes, processes with ephemeral stochastic templets. In contrast, biological patterns are determinate patterns, and the uniquely biological templets have stability, coherence, and permanence. Random patterns are the complex stochastic patterns, but biological patterns are the complex determinate patterns. Stable templets – reproducibility – was the great leap, for life is matter that learned to recreate faithfully what are in all other respects random patterns.

8

Reductionism

Reduction... is the explanation of a theory or a set of
experimental laws established in one area of inquiry, by a theory
usually though not invariably formulated for some other
domain.
[E. Nagel (1961) *The Structure Of Science. Problems In The
Logic Of Scientific Explanation* Harcourt, Brace & World NY,
p. 33.8.]

Science strives to reduce our experiences to symbols. Experiences
are colorful, multi-faceted, and fuzzy along the edges; symbols
are bland, one-dimensional, and precisely-bounded. Real world
observations can be bulky and ill-shaped and can have both
strong and tenuous ties with a myriad of other real world observations; abstractions are built of simple, smooth-surfaced elements,
uncoupled from other constructs, and abstractions can easily be
carried in one's pocket. Scientifically, we give up the shifting and
elusive mystery of the world, but, in exchange, we gain the
standardized and reproducible abstractions from which we can build
precise determinate explanations.

The reduction of experience to useful symbols – the construction
of scientific abstractions – is the essential scientific endeavor.
In this sense, all of science is reductionism. At the same time, there
is another sense to scientific reductionism. Once a real world
phenomenon has been abstracted, the scientist attempts to create a
scientific explanation of that abstraction, and this too is a type of
reductionism. To reduce is to recreate, and reductionism is actually
constructionism.

A scientific abstraction is a model of some real world phenomenon,
and a scientific explanation is then a further transformation of that
phenomenon, carried out largely in the abstract realm. Scientific
explanations break the original abstraction into certain prescribed
classes of parts: in a scientific explanation, the original abstraction

is reduced to a different, more elemental set of abstractions – predecessor abstractions. In addition, a scientific explanation defines the particular relationships between the parts that are necessary to reproduce the whole.

The reduction of an abstraction to a scientific explanation is a second level of scientific reductionism. Is this a sufficient depth of reduction for scientific understanding? Or, must the scientist reduce phenomena still farther before he can step back, survey his analytic constructs, and feel that he has made some sense of the real world? Is there a natural end to scientific reduction, or must it be carried on forever?

In practice, reductionism usually settles at a natural stopping point. Consider DNA. DNA is the molecule of inheritance: most of the information passed from generation to generation in the lineages of organisms is encoded in the specific subunit sequences of DNA molecules. The biologist considers DNA a sufficient explanation for the inheritance of protein, and he is satisfied to explain particular inherited features in terms of lists of DNA sequences. Of course, the biologist could further reduce his explanation to the level of quantum mechanics, but he does not feel the need. A quantum mechanical description of inheritance represents more reduction than is necessary. Reduction to the level of DNA is reduction to the level of natural explanatory units.

What constitutes a natural explanatory unit? A well-constructed configurational explanation fulfills R. Carnap's definition of a scientific explanation [(1966) *Philosophical Foundations Of Physics* Basic Books, NY], and it is in some sense a whole and natural explanatory unit and thereby an acceptable level of scientific reduction. A configurational explanation is a formalism in which a pattern is replaced by the logical conjunction of two predecessor patterns: a set of precursor elements and a templet; and, in this way, reduction to a configurational explanation is the translation of one pattern into different, independent patterns. The predecessor patterns, the set of precursors and the templet, contain all of the necessary information for building the final pattern, and, together with the machinery of a configurational explanation, the precursors and the templet are entirely self-sufficient. The configurational explanation is a self-sufficient isolated system that completely replaces the original pattern. Further reduction can only replace this new system with yet another isolated system, but further reduction will remove us farther from the original pattern. By constructing a thoughtful configurational explanation, the scientist has reduced the organization – the form – of a real world pattern to an explanatory level that is as close to the original pattern as possible. Reduction to more distant explanations dims our understanding and cheats our metaphysical purpose:

... the analysis of the hierarchy of living things shows that to reduce hierarchy to ultimate particulars is to wipe out our very sight of it.

[M. Polanyi (1968) 'Life's irreducible structure' *Science* **160**: 1312.]

Reduction of a nonanalytic expression: the pendulum and the robot

Configurational explanations are especially useful in putting scientific form upon those phenomena that feel intractably complex, phenomena such as nonanalytic expressions. 'Nonanalytic expression' is a general label for phenomena that cannot be summarized in simple formulas. Consider, for instance, two machines, B and B', that manufacture binary strings A and A' each time the machines are turned on:

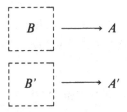

Each machine contains an ideal pendulum, and the output string is built in discrete 'ticks'. When the pendulum swings to the right, it causes a one to be output; when the pendulum swings to the left, it cause a zero to be output.

The first machine, machine B, contains only a pendulum, and its output A will always be 10101010101010101010101010... (Fig. 2). The second machine, machine B', also contains a robot and a specific internal instruction manual, another binary string, $C = 11101101101111101000000010000....$ In B', the pendulum is

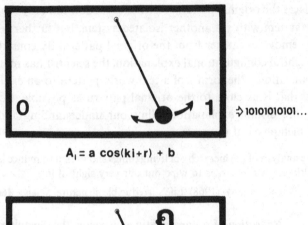

$$A_i = a \cos(ki+r) + b$$

$$A_i = \text{nonanalytic}$$

Fig. 2

not free to swing; instead, its robot moves the pendulum according to the instructions. The output A' of machine B' is always completely determined by the internal string C (Fig. 2).

To build configurational explanations for each of these output patterns, we model our machines B and B' as logical adders LA and LA':

$$MV \longrightarrow \boxed{LA} \longrightarrow A$$
$$TV \longrightarrow$$

$$MV' \longrightarrow \boxed{LA'} \longrightarrow A'$$
$$TV' \longrightarrow$$

For machine B the right and the left swings of the pendulum – the precursor elements – can come in only one sequence; thus, the underlying 'universal law' is limiting, the raw materials are topologically knowledgeable, and the matching vector, MV, is very restrictive. Specifically, MV has a minimum number of ones, and, for an output string of length n, the minimal necessary templet size is $S = 2(n - 1)$ and the templeting index $Q = 2(n - 1)/n^2$. In other words, the longer the output string, the more it can be considered self-assembling and the simpler it will be.

In contrast, the raw materials for machine B' can best be modelled as topologically naive – because the pendulum is actually robot-driven, the right and the left movements of the pendulum can occur in any possible sequence – and the matching vector in our configurational explanation of the output pattern A' should have a maximum number of ones. For an output string of length n, $S = 2n(n - 1)$ and $Q = 2(n - 1)/n$; in other words, regardless of its length, the output string is always almost entirely templeted and it can be considered to be maximally complex. Thus, the pattern A is much simpler than the pattern A'.

Pattern A can be fully summarized by its 'universal law', a formula – the formula for a simple harmonic oscillator – and, in this way, automaton LA models the formula:

$$A_i = a(\cos{(ki + r)}) + b$$

Through our configurational explanation, the general formula for a simple harmonic oscillator can now be assigned a complexity value, namely: $Q = 2(n - 1)/n^2$ (where n is the length of A).

In theory, pattern A' can also be summarized by a formula, but in this case the formula cannot be written in a very brief form and it would have to be termed 'nonanalytic'. A simple description of the machine B' with an explicit listing of the internal string C would probably be about as brief a formula as is possible. Once again, our configurational explanation models this formula and allows us to assign it a complexity value, namely: $Q = 2(n - 1)/n$. As might be expected, a nonanalytic expression has essentially maximal complexity.

Pattern A can be reduced to a simple formula. This type of reduction – reduction to simpler predecessors – is the usual type of scientific reduction, and in the physical sciences it is often assumed to be the 'right' type of scientific explanation. On the other hand, some natural phenomena cannot be reduced to simpler predecessors; the prime examples are the complex biological patterns. These patterns are like the outputs of machine B' – they are nonanalytic expressions which can only be reduced to predecessors that are as large and as intricate as the final patterns themselves.

Templets

Although it is most satisfying to reduce to smaller, simpler and less intricate abstractions, smaller, simpler and less intricate are not necessary. For some natural phenomena, there simply is no reduction to smaller predecessors. In these cases, the companion rule to 'order stems from order' is that 'complexity stems from complexity'. For analyses of the form and the organization of phenomena, configurational explanations offer a standard reductionist paradigm unconstrained by the demands of reduction to simplicity. Here, templets are the common currency, the standard forms with which precisely to describe and to compare the transformations in order, in form, and in complexity.

The satisfaction inherent in simple explanations is a characteristically human feature, arising from the difficulty that our brains have in remembering and in manipulating large, multipartite, and oddly-shaped abstractions. True, we are adept at testing general relations between big and amorphous abstractions, but we do poorly at working out the precise and logical details of their interrelations. Computers have the opposite constraints: they are adept at precise and logical manipulations of abstractions regardless of size, but they do poorly when forming general, imprecise, and intuitive relations between abstractions. Were computers to have written the rules for scientific explanation, 'small' and 'simple' would not have appeared.

Reductionism is more than the translation of one set of abstractions into another: it is a translation that gives the scientist insight into natural relations. For this reason, the reductionist translation must mirror an actual transformation found in the natural world. When explaining the form or the organization or the order of a natural pattern, the abstractions must also model the natural transformations. Configurational reductionism explains the form of a pattern in terms of its transformation from prescribed predecessor forms, and configurational explanations must match their abstract forms to appropriate real world counterparts.

The appropriate real world templet may be simple or it may be complex, but in all cases it is the central element in a reductionist configurational explanation. The templet is an explicit carrier of form, and the abstract templet always corresponds to substantive entities in the real world. Form, organization, and order are not ethereal properties floating in the vapors of the natural world to suddenly descend and coalesce upon real world objects. Form, organization and order, like mass, energy and entropy, exist only in relation to substance – to matter. They are integral properties of material patterns, and they are woven into assemblages of real world items.

Like mass, energy and entropy, form, organization and order obey laws of conservation. Order cannot be created *ex nihilo*. Order is inherited from preexisting order, and it is perpetuated through templets. When the templets are ephemeral, impermanent, and unstable – when the pattern-assembly processes are stochastic – the detailed order seems elusive to us. When the templets have a permanence, the patterns can be recurrent and the order seems fairly solid. In either case, whether the templets are unstable or stable, ephemeral or permanent, the order of the universe is nonetheless always propagated through a sequence of templets – where templets beget templets beget templets, and on forever.

The arrow of templeting

Complex processes often have an arrow, and this polarity limits their reduction. Stochastic processes are usually polarized – it is hard to uniquely identify the predecessors in a stochastic process, and thus it is difficult to produce a reductionist explanation of a random pattern. Entropy is responsible for the polarity of natural processes:

Idealized processes may be reversible and proceed indefinitely in the forward or backward directions, but all natural processes proceed in only one direction, toward an equilibrium state, and upon reaching this state they come to a halt.

[A. Katchalsky & P.R. Curran (1967) *Nonequilibrium Thermodynamics In Biophysics* Harvard Univ. Press Cambridge, MA, p. 10.]

So far as physics is concerned time's arrow is a property of entropy alone.

[A.S. Eddington (1929) *The Nature Of The Physical World* Macmillan, NY, p. 80.]

Entropy is the degree of disorganization of a system, and, in the most general terms, entropy EN is defined as:

$$EN = k \log N$$

where N is the number of different realizable states and k is a constant. Expressed in this form, entropy is a measure of information; in a system of high entropy, much information is needed to specify one particular state. The information I of a particular state is defined as:

$$I = -k' \log p(N)$$

where $p(N)$ is the probability of occurrence of the particular state and k' is a constant, and:

$$p(N) = 1/N$$

Thus:

$$I = (k'/k)EN$$

– the information needed to specify a particular state is directly proportional to the entropy of the entire system.

In terms of configurational explanations, N is the number of distinct possible configurations that can be constructed from a given set of precursor elements. S, the minimal necessary templet size for a given configurational explanation (also, the informational content of that configurational explanation), is:

$$S = K \log N$$

where K is a constant. Hence:

$$EN = (k/K)S$$

S equals the number of ones in the matching matrix MM, thus:

$$EN = (k/K)\left(\sum_{\text{row}} MM \right)$$

The entropy of a pattern-assembly process is directly proportional to the extent of the interactive potentials of the precursor elements and to the restrictiveness of the underlying 'universal assembly laws'. Topologically knowledgeable raw materials generate low entropy systems, and topologically naive raw materials generate high entropy systems. Low entropy systems produce minimally complex patterns, and high entropy systems produce maximally complex patterns.

The general polarity of stochastic processes is unchallenged, but the overlap of stochastic processes and complex pattern-assembly processes can confuse the casual theoretician. Time's arrow – the polarity of natural processes – points toward states of higher entropy, and this is usually said to mean states of greater disorganization. This phraseology has led to a fundamental misunderstanding about certain natural pattern-assembly processes that fabricate complex patterns. To expose the underlying problem, let me begin with the complex stochastic patterns.

Complex patterns require maximal templets, and individual complex patterns are members of a large set of equivalent possible patterns. Besides needing complete templets, the complex stochastic patterns – random patterns – derive from stochastic pattern-assembly processes, processes for which the real templets are ephemeral and stochastic. (See the Appendix.)

The production of a random pattern clearly follows the entropically appropriate directional flow of natural events. Events roll toward disorder: the production of randomness is the production of maximally complex and unpredictable patterns. The products are always at least as complex as their predecessors, and there is no violation of the law: 'Entropy tends to remain constant or to increase.'

In contrast, consider the determinate production of a complex pattern, such as a person. Here, there has always been some haziness as to how to smoothly mesh our definitions of entropy, complexity, order, and directional events. In biology, the direction of natural events seems clear: the arrow points toward the production of the final biological pattern, the person. A problem arises, however, from our feeling that people represent highly ordered patterns (complex though that order is) and that in biology the natural flow of events is – contrary to the second law of thermodynamics – in the direction of creating more order rather than more disorder.

The apparent paradox is usually resolved by considering that although an organism may become more ordered – an apparently unnatural direction of spontaneous events – the whole system, an organism plus its environment, always becomes more disordered – the natural direction of spontaneous events. As a mathematical physicist has written:

Order, organization, is the characteristic of life. Hence, the impression that life, in its evolution on earth, resists the plunge into the abyss of 'thermal death', which the entropy law imposes upon inorganic matter. Bergson has coined the grandiose words '*elan vital*' for this resisting power. Even a crystal, as it grows, creates order in the substance which it seizes. When water crystallizes into snow or ice it gives off heat to its environment, that is, the entropy of the environment, its 'disorder', increases. Thus, *in toto*, the entropy law remains intact. Similarly an organism which grows up from the fertilized egg by progressive differentiation has the capacity to create order – at the expense of the environment, to be sure, whose entropy, or disorder, increases accordingly.

[H. Weyl (1949) *Philosophy Of Mathematics And Natural Science* Princeton Univ. Press, Princeton, p. 212.]

Or, as expressed by a biologist:

The puzzle is that living organisms are very highly ordered, at every level. Order is strikingly apparent in large structures such as a butterfly wing or an octopus eye, in subcellular structures such as a mitochondrion or a cilium, and in the shape and arrangement of molecules from which these structures are built. The large number of atoms in each molecule of protein or nucleic acid have been captured, ultimately, from their highly disorganized state in the environment and locked together into a precise structure. Every time large molecules are made from smaller ones, as when a living cell grows, order is created out of chaos. Even a nondividing cell requires constant ordering or repair processes for survival, since all of its organized structures are subject to spontaneous accidents and side reactions. How is this possible thermodynamically? We shall see that the answer lies in the fact that the cell is constantly releasing heat to its environment, and therefore it is not an isolated system in the thermodynamic sense.

[B. Alberts, D. Bray, J. Lewis, M. Raff, K. Roberts & J.D. Watson (1983) *Molecular Biology Of The Cell* Garland, NY, p. 62.]

Although it is undoubtedly true that for most biological patterns the entropy of the overall system does increase, the descriptor 'ordered' only serves to confuse the issue. 'Order' is loosely used in contrast to 'random', but we have seen that biological ordered patterns share a critical characteristic of random patterns – both are maximally complex. In actuality, the production of a biological pattern differs from the production of a random pattern in but one respect: for biological patterns – complex determinate patterns – the real templets are relatively permanent, while for random patterns – complex stochastic patterns – the real templets are ephemeral and stochastic. A person differs from a random pattern only in the reproducibility of its details. The concept of 'ordered' as loosely applied to complex patterns actually means 'predictable' or 'matchable to a previously extant pattern', and those aspects of people that strike us as ordered could better be called recurrent or stereotyped. The wonder of life is not in its apparent creation of 'order'. Rather, this wonder is in its stereotyped assembly of very complex patterns. It is directly in the stability and permanence of life's templets that the natural entropic forces are resisted.

Templeting and the creative act

Templeting processes are indigenous to the natural realm; templets are the natural basis for any number of real world patterns. Nonetheless, in the natural world, templeting shares the pattern-assembly chores with self-assembly, and both processes contribute to the form of our natural universe. The spiritual world, on the other hand, is dominated by templeting.

Artistic creations are, in essence, entirely templeted. The heart of a painting, a sculpture, a poem, a concerto is the painter, the sculptor, the poet, the composer. Each artwork is fabricated of topologically naive elements, elements that can equally well be assembled in an almost limitless number of distinct patterns. The underlying 'universal assembly laws' are unrestrictive, and it is the artist who explicitly determines the particular form of the artwork. Working in a medium of unrestricted potential, an artist fashions one particular final pattern, and in this way art is always completely templeted.

Natural patterns can either be templeted or self-assembled. Artistic patterns must be templeted. Truly artistic patterns cannot be summarized, abbreviated, or condensed; they cannot be fully reduced to smaller, simpler, or less intricate components. The creative act is synonymous with templeting. It is this fact that contributes to the discomfort that a scientist feels when ascribing a natural pattern to a templeting process. It is this fact that colors our feelings about any templeting process and that, for example, adds a special aura to the most well known of all templeting acts:

And there the great God Almighty
Who lit the sun and fixed it in the sky,
Who flung the stars to the most far corner of the night,
Who rounded the earth in the middle of His hand;
This Great God,
Like a mammy bending over her baby,
Kneeled down in the dust
Toiling over a lump of clay
Till He shaped it in His own image
[James Weldon Johnson 'The Creation']

Conclusions

Scientists collect data – they carry out a natural historical calatoguing of the real world – and then they interweave the data in a synthetic constructure. In this way, a scientist builds the theoretical basis for his particular discipline.

To construct a synthetic science, the scientist reduces: natural observations are reduced to scientific abstractions, and these abstractions can sometimes be smaller, simpler, and less intricate than the real phenomena; this has often been the case in physics. At other times, the natural phenomena cannot be reduced to simpler predecessors; this is often the case in biology. Nonetheless, both disciplines need a uniform reductionist paradigm, and, in terms of the form, the order, and the organization inherent in any phenomena, configurational explanations represent one general reductionist paradigm.

Configurational explanations are abstract isolated systems that can replace and can thereby explain particular patterns. Inside a configurational explanation, the standard predecessors are the precursor elements and the templet. Specific order is inherited through templets, and it is through templets that comparisons of the organization of apparently disparate forms can be made.

For any given pattern, the set of precursor elements and the underlying 'universal laws' are unique. On the other hand, the particular situational constraints, the real templet, can often be any one of a number of distinct but equivalent templets. In the natural realm, the number of equivalent templets is not a polarizing factor for pattern-assembly – stochasticism is the directional force in the real world. If the pattern-assembly process is largely determinate, then the real world process itself has little intrinsic polarity regardless of the number of potential templets. In the abstract realm, however, templeting gives an arrow to scientific explanations. If the real templet is a member of a large class of equivalent templets, then the abstract explanation is strongly polarized. The scientist cannot reverse the direction of the explanation and peer into it to discover which templet has actually been at work. The fact that Nature need not behave parsimoniously, that natural patterns can be built with excess or unnecessary complexity, makes *a priori* reasoning and abstract inference insufficient to uniquely reduce those phenomena that do not require complete templets. Self-assembling patterns cannot be uniquely reduced by the armchair theoretician.

Artistic creations are not self-assembling; they are largely templeted. Being constructed of topologically naive raw materials, artistic creations require maximal templets. Maximal templets are unique, and thus explanations of artistic creations are not polarized. An artistic creation can be uniquely reduced, and the mystery of creativity does not reside in its intransigence to reduction. Instead, the creative act is characterized by its irreducible complexity. The templets of the creator must be as large and as intricate as the creation; although it can be reduced, art reduces to an explicit image of itself. 'What can be explained is not poetry', [C. Sandburg (1930) *Early Moon* Junior Literary Guild, NY, p. 17], and what can be reduced to smaller, simpler, and less intricate beginnings is not art.

Appendix: A pattern theoretic formalization

A formalized scientific abstraction

Scientific abstractions should be built in precise, symbolic, and formal language. Formality imposes a well-organized structure on long or convoluted logic statements, and the translation of an argument into formal language is often the only way to guarantee that it is the logic that is convincing and not merely the rhetoric.

For discrete abstractions in general, Ulf Grenander has developed a very useful formal language, which is an almost literal translation of the narrative description presented in this book. [U. Grenander: (1976) *Pattern Synthesis*, (1978) *Pattern Analysis*, (1981) *Regular Structures*; Springer-Verlag, NY.] Basically, Grenander provides a molecular model in which the pattern elements are atoms called 'generators' and the topological interconnections among elements are binary relations manifested by the presence or the absence of links between the 'bonds' of each generator. An entire abstraction, composed of many interconnected generators, is then a multi-dimensional molecule or, as mathematicians call it, a graph.

Generators

The elements of a scientific abstraction are generators (Fig. 3a). Each generator, g, is a discrete atom characterized by a definite and finite set of attributes; i.e., $g = \{a_1, a_2, \ldots, a_n\}$. Generators are the nodes in the molecular spiderweb forming a scientific abstraction, and they compose the substance of the abstraction.

Bonds

Each generator has a fixed and finite set of bonds by which it can be interconnected to other generators. A generator with two

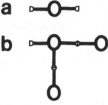

Fig. 3

bonds can by symbolized as $- O -$ (Fig. 3a). Two bonds that are linked form an interconnection between generators, and this interconnection is a determinate binary phenomenon – the bonds are either definitely connected, or they are definitely not connected. Linked bonds form the scaffolding in the molecular model of a scientific abstraction (Fig. 3b).

Matching relations and the matching matrix

The aim of science is not things themselves,... but the relations between things; outside those relations there is no reality knowable.

[H. Poincare (1952) *Science And Hypothesis* Dover, NY, p. *xxiv.*]

The necessary condition for any two bonds to be linked is that they fulfill a prespecified matching relation, m. This condition is necessary but not sufficient – beyond fulfilling m, the bonds must also be given the opportunity to match by being brought into appropriate proximity.

Matching is a binary relation. Therefore, all of the potential bond couples can be simply categorized as either permissible (regular) or not permissible (irregular). In most cases, permissible linkages can be summarized in a table of generators, called a matching matrix. The matching matrix is a binary matrix, MM; for example:

$$MM = \begin{array}{c} \\ a \\ b \\ c \\ d \\ e \end{array} \begin{array}{ccccc} a & b & c & d & e \\ 0 & 1 & 0 & 0 & 0 \\ 1 & 0 & 1 & 0 & 0 \\ 0 & 1 & 0 & 1 & 0 \\ 0 & 0 & 1 & 0 & 1 \\ 0 & 0 & 0 & 1 & 0 \end{array}$$

where a one in the ith row and the jth column means that generators i and j can be interconnected and a zero in that position means that the corresponding generators cannot be interconnected.

This particular matching matrix indicates that, due to the inherent nature of their bonds, the generators {*abcde*} can be connected into strings in the order *a-b-c-d-e* but not in any other order. If *a-b* symbolizes an interconnection between the generators *a* and *b*, then:

$$a\text{-}b \quad b\text{-}c \quad c\text{-}d \quad d\text{-}e \quad a\text{-}b\text{-}c \quad b\text{-}c\text{-}d$$
$$c\text{-}d\text{-}e \quad a\text{-}b\text{-}c\text{-}d \quad b\text{-}c\text{-}d\text{-}e \quad a\text{-}b\text{-}c\text{-}d\text{-}e$$

are permissible strings of generators, but:

$$a\text{-}c \quad b\text{-}d \quad c\text{-}e \quad d\text{-}a \quad a\text{-}c\text{-}d \quad b\text{-}d\text{-}e$$
$$c\text{-}e\text{-}a \quad a\text{-}d\text{-}e\text{-}c \quad b\text{-}c\text{-}e\text{-}d \quad a\text{-}b\text{-}c\text{-}e\text{-}d$$

are not permissible strings.

The matching matrix summarizes the 'universal assembly laws' governing a set of generators.

Topology and the adjacency matrix

The actual interconnections that have been formed between generators define the neighbor relations of a scientific abstraction. In this way, the multidimensional bond linkages – the spiderweb of scaffolding – is an explicit representation of the topology of the abstraction. Two generators that are interconnected are topological neighbors; and, regardless of how close together one might picture them to be, if two generators are not explicitly interconnected, then they are not topological neighbors.

One useful representation of a scientific abstraction is a graph schematic, with the generators drawn as vertices and the interconnections as edges (Fig. 3b). Another important representation is as a binary matrix, an adjacency matrix. Here, the generators are the row and column identifiers and the extant interconnections are entered as ones in the appropriate matrix positions. For example, the adjacency matrix, AM:

		a	b	c	d	e
	a	0	1	0	0	0
	b	1	0	1	0	0
$AM =$	c	0	1	0	1	0
	d	0	0	1	0	1
	e	0	0	0	1	0

represents the scientific abstraction

$$a\text{-}b\text{-}c\text{-}d\text{-}e$$

The adjacency matrix for a scientific abstraction is a precise and complete summary of the topology of that abstraction.

Configurational explanations

A configurational explanation describes the basis for the internal organization of a pattern – the order within a pattern. The internal organization or the topology of a pattern arises from the interplay of two sets of topological forces. First, there are the intrinsic topological constraints of the individual elements of the pattern. Due to their particular interactive natures, the precursor elements can be assembled in only certain configurations. Formally, a predecessor abstraction – the local precursor abstraction – summarizes all of the potential interconnections between the elements of a pattern. This abstraction is termed 'local' because it disregards long-range, secondary, or synergistic interactions. The local precursor abstraction is concerned only with potential and permissible interactions between immediate neighbors. It represents the non-teleological, self-assembling forces – the 'universal laws' – of pattern assembly.

In contrast, the templet of the configurational explanation is the formal embodiment of the additional topological constraints that must be imposed to produce one particular final pattern of interest. Templets explicitly describe the specific subset of opportunities that must be provided if the elements are to coalesce into just the one required configuration. If pattern-assembly is thought of as being carried out by a machine, then the templet is the set of instructions describing which precursor elements should be given the opportunity to interconnect during the assembly process.

A configurational explanation can be symbolized as the logical addition of two matrices (Fig. 4):

$$MM \wedge T = AM$$

where $\wedge$ is the commutative and associative matrix operation AND (logical conjunction), MM is the matching matrix derived from the local precursor abstraction, AM is the adjacency matrix derived from the final abstraction (the scientific abstraction of the pattern itself), and T is the binary matrix, the templet, specifying the additional topological information that is necessary to assemble the one particular pattern of interest. Templets are the heart of configurational explanations, and templets are the standard currency of form.

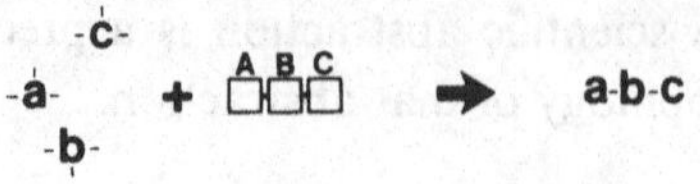

$$
\begin{array}{c} \text{-c-} \\ \text{-a-} \\ \text{-b-} \end{array} \quad + \quad \overset{\text{A B C}}{\square\square\square} \quad \Rightarrow \quad \text{a-b-c}
$$

$$
\begin{matrix} 011 \\ 101 \\ 110 \end{matrix} \quad \wedge \quad \begin{matrix} 010 \\ 101 \\ 010 \end{matrix} \quad = \quad \begin{matrix} 010 \\ 101 \\ 010 \end{matrix}
$$

$$
\mathsf{M} \;\wedge\; \mathsf{T} \;=\; \mathsf{A}
$$

Fig. 4

By combining a set of precursors with a templet, a configurational explanation explicitly describes the creation of the organization or the topology of a pattern. To further define the basis for the pattern, an isolated configurational explanation should be set in its historical context. In this way, the lineages of the precursors and of the templet can also be made explicit, and one can also see the role of the pattern in determining subsequent patterns.

To make precise the contextual basis for a pattern and to make explicit all of the necessary internal operations, configurational explanations can be modelled as finite automata. The constructs of Pattern Theory are closely allied with the constructs of Automata Theory, and to translate configurational explanations into standard automata, the matrices are strung out as binary vectors:

$$
MV \wedge TV = AV
$$

where MV is the matching vector made from the matching matrix MM, TV is the templet vector made from T, and AV is the adjacency vector made from AM. The configurational explanation is now equivalent to a finite automaton, a logical adder LA, that will logically add any two binary vectors:

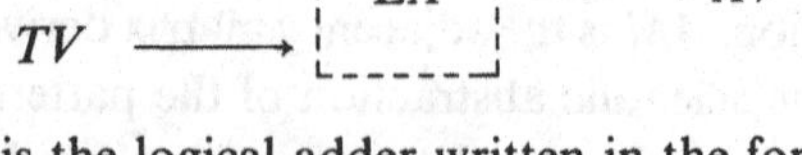

$$
\begin{array}{ccc} MV & \longrightarrow & \\ & & \boxed{LA} \longrightarrow AV \\ TV & \longrightarrow & \end{array}
$$

Here is the logical adder written in the form of a brief computer program. (This exercise – the translation of an operational concept into an explicit algorithm – helps to bare any assumptions quietly hidden behind the words of a narrative description.) In BASIC:

```
10  PRINT '**LOGICAL ADDER**': PRINT
20  PRINT '   MV, TV AV'
30  INPUT; '      ', MV, TV
40  AV = MV AND TV: PRINT '      '; AV
50  GOTO 30
```

produces the following printout when operating on the *MV* vector 011101110 and the *TV* vector 010101010:

```
**LOGICAL ADDER**
    MV, TV   AV
    0, 0      0
    1, 1      1
    1, 0      0
    1, 1      1
    0, 0      0
    1, 1      1
    1, 0      0
    1, 1      1
    0, 0      0
```

In this algorithm, the two input vectors are read simultaneously digit by digit and the output vector is also created digit by digit. There is no limit to the length of either the input vectors or the output vector, and there are no special codes for the beginning or for the end of the vectors. An outside agent must ensure that the two input vectors are appropriately aligned when they are first fed into the computer; after this, the computer runs autonomously.

As automata, configurational explanations can easily be combined into more complex functions that will model concatenated explanations – certain scientific theories. For example, logical adders can be concatenated in parallel as:

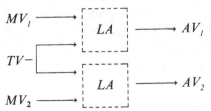

With such representations, various higher level phenomena can be analyzed as scientific theories (concatenations of scientific explanations). For example, consider a simple scientific theory that is reduced to a set of stacked (parallel) explanations. This can be modelled as a set of N logical adders in parallel, each with the same templet vector. What is the probability that all of the outputs would accurately reproduce a certain predetermined pattern, AV_2? If the common templet is chosen randomly, then the probability that any one logical adder would output one particular vector is:

$$p = 2^{(-QS_{max})}$$

where Q is the templeting index and S is the size of the adjacency vector. The probability that N logical adders would all have appropriate outputs is the product of all the individual probabilities; i.e.:

$$p^N = 2^{(-NQS_{max})}$$

This probability decreases dramatically as the length of the vectors increases (represented by S_{max}), as the number of ones in the matching vectors increases (represented by Q), and as the number, N, of logical adders increases. Large sets of parallel logical adders, logical adders processing long vectors, and logical adders producing templeted outputs, all require highly controlled, non-random inputs to operate accurately. Under these conditions, reproducible ordered outputs require a great deal of preexisting order. Large stacked scientific theories are not self-correcting and require detailed, accurate, and carefully monitored inputs to produce accurate outputs.

By interconnecting simple two-state automata it is possible to simulate any finite automaton and to model a wide variety of scientific theories, but in general one must build such simulations using other varieties of two-state automata in addition to the logical adder. (For a good discussion of this problem, see M.L. Minsky (1967) *Computation: Finite And Infinite Machines*, Prentice-Hall, chapter 3.) Specifically, two other two-state automata are needed to build universal finite automata: disjunction automata, and A AND NOT B automata.

Disjunction automata

Consider a pattern made of the four cities: London, New York, Chicago, and Los Angeles, represented by the four generators L, NY, C, and LA. Now, imagine that the Local Railroad Company plans to interconnect these cities by a set of ralways: New York to Chicago and Chicago to Los Angeles. London, of course, will remain unconnected by rail.

A configurational explanation for the proposed railway pattern is as follows. The matching matrix is:

	L	NY	C	LA
L	0	0	0	0
NY	0	0	1	1
C	0	1	0	1
LA	0	1	1	0

– the three continental cities can all potentially be directly interconnected by railways. The adjacency matrix is:

	L	NY	C	LA
L	0	0	0	0
NY	0	0	1	0
C	0	1	0	1
LA	0	0	1	0

representing the actual pattern of interconnections:

$$L \quad NY-C-LA$$

The templet matrix is:

	L	NY	C	LA
L	0	0	0	0
NY	0	0	1	0
C	0	1	0	1
LA	0	0	1	0

– this matrix represents the master railway blueprint of the Local Railroad Company.

Concurrently, another firm, the Express Railroad Company, has a different blueprint. The Express Railroad Company plans to directly interconnect New York and Los Angeles, bypassing Chicago. The

Express Railroad Company operates with the same matching matrix as does the Local Railroad Company, because the precursor elements are identical. On the other hand, the Express Railroad blueprint is represented by the templet matrix:

	L	NY	C	LA
L	0	0	0	0
NY	0	0	0	1
C	0	0	0	0
LA	0	1	0	0

and, here, the proposed railway pattern is:

$$L \quad \underbrace{NY \quad C \quad LA}$$

Economic masterminds set to work, the two companies merge, and the new E & L Railroad Company now forges a combined blueprint for intercity railways, merging the two original blueprints. Eventually, the new railway connections will be:

$$L \quad \underbrace{NY - C - LA}$$

and this final rail pattern is based on the combined templet matrices:

	L	NY	C	LA
L	0	0	0	0
NY	0	0	1	1
C	0	1	0	1
LA	0	1	1	0

The templets have been easily and naturally merged by forming their union. (Notice, however, that it does not make sense to combine matching matrices in the same simple manner. The given cities either can or cannot be connected by rail, and to change the matching matrix is equivalent to changing the basic properties of the cities – i.e., to redefining the precursor elements and their 'universal assembly laws'.)

The two original templets have been combined by logical disjunction, or logical OR, symbolized by $\vee$, and this process is carried out by a disjunction automaton, DA:

$$TV_1 \longrightarrow \boxed{DA} \longrightarrow TV_3$$
$$TV_2 \longrightarrow$$

A AND NOT B automata

Together, and when concatenated in sufficient numbers, logical adders and disjunction automata can model a variety of more complex machines. There are, however, a great many automated functions that cannot be built with just those two types of elements. To model any possible finite machine, one needs an additional type of element, an *A* AND NOT *B* automaton. With such an automaton in hand, the range of possible automated functions expands tremendously.

The *A* AND NOT *B* automaton takes two inputs, *A* and *B*, and produces a single output that is the logical addition of *A* with the negative of *B*. In the language of pattern operations, this automaton corresponds to a matrix maneuver that identifies irregular locations in the final pattern. Irregular locations are places where the generators do not fit properly.

Through a simple matrix operation, the templet matrix and the matching matrix can be used to identify irregular locations in the pattern, sites of instability in the overall pattern. By taking the negative image of the matching matrix and then logically adding it to the templet matrix, a binary irregularity matrix is produced in which ones signify those sites where two generators are effectively adjacent but do not fit; i.e.:

$$T \wedge \sim MM = IM$$

where $\sim MM$ denotes the negation of the matching matrix (ones replaced by zeros, and zeros replaced by ones), and IM is the matrix of irregular connections – an irregularity matrix. When written in terms of vectors, this becomes:

$$TV \wedge \sim MV = AV$$

which symbolizes the *A* AND NOT *B* automaton:

What is the meaning of an *A* AND NOT *B* automaton in the context of configurational explanations? An *A* AND NOT *B* automaton is a device that computes an irregularity matrix for a particular adjacency matrix. The irregularity matrix represents an image of the rough edges, the ill-fitting areas, and the unstable locales in a given pattern. Most importantly, it reveals the impermanent, ephemeral adjacencies in the pattern.

Together, logical adders, disjunction automata, and *A* AND NOT *B* automata can model a vast array of scientific theories. In a number of cases, such concatenated automata can be simplified and replaced by single logical adders. This is equivalent to simplifying certain scientific theories by reducing them to individual scientific explanations. For example, a string of N logical adders in series can be replaced by one logical adder, where the effective matching vector is actually the logical sum of the N internal matching vectors: scientific theories that are strings of explanations can be reduced to individual explanations. The concatenation of a disjunction automaton and a logical adder can also be replaced by one logical adder, where the effective templet vector is actually the logical disjunction of the internal templet vectors: scientific theories of serial AND–OR explanations can be reduced to individual OR explanations.

Counting configurations

A matching matrix represents the 'universal laws' inherent in a configurational explanation, and many of the characterizations of a pattern assembly system depend on counting the number of configurations that are possible under a given matching matrix. For small matching matrices, in which all of the precursor elements of a given pattern can be listed explicitly, this number can be derived by counting or by various practical algorithms. For very large matching matrices, the calculation can be more difficult.

When the number, n, of generators is very large, it is convenient to encode the matching information in a more condensed form. In condensed matching matrices, sets of generators that are indistinguishable are summarized by only a single row of values. For example, the matching matrix:

$$
M = \begin{array}{c c c c c}
 & a & b & b & c \\
a & 0 & 1 & 1 & 0 \\
b & 1 & 0 & 1 & 1 \\
b & 1 & 1 & 0 & 1 \\
c & 0 & 1 & 1 & 0 \\
\end{array}
$$

can be condensed into a binary matrix:

$$M = \begin{array}{c} \\ a \\ b \\ c \end{array} \begin{array}{ccc} a & b & c \\ 0 & 1 & 0 \\ 1 & 1 & 1 \\ 0 & 1 & 0 \end{array}$$

U. Grenander and Y.-S. Chow have pointed out that the number of patterns permitted by a condensed matching matrix can be estimated from its spectral radius. Specifically, it appears that the number N_n of possible patterns is asymptotically related to the spectral radius r_1 of the underlying matching matrix as:

$$\lim_{n \to \infty} ((1/n)\log N_n) = \log r_1 \tag{1}$$

I will briefly demonstrate this relationship with simple string patterns.

Begin with an unlimited number of g distinct types of generators, each having only two bonds. Such generators can form patterns that are either strings or rings. Here, the number N_n of regular configurations of size n is:

$$N_n = \sum_{i_1, i_2, \dots i_n} m(i_1, i_2)m(i_2, i_3) \dots m(i_{n-1}, i_n)$$

where $i_1, i_2, \dots i_n$ represents any set of n integers all less than or equal to g and the summation is over all such sets. In terms of a matrix power, this can be more simply expressed as:

$$N_n = eM^n e'$$

where e' is the $n \times 1$ column vector $(1, 1, \dots, 1)$. [See also the related graph theoretic calculation of the number of walks of length n in the matrix representation of a graph – Theorem 13.1, F. Harary (1969) *Graph Theory* Addison-Wesley, Reading MA, p. 151.] For this expression and for symmetric matrices M, there always exists a matrix U such that $M = UJU^{-1}$ and the Jordan decomposition of M^n yields $M^n = UJ^n U^{-1}$. J can be summarized in the form:

$$J = \begin{vmatrix} r_1 d_1 & 0 & \dots & \dots & 0 \\ 0 & r_2 d_2 & 0 & \dots & 0 \\ \dots & \dots & \dots & \dots & \dots \\ \dots & \dots & \dots & \dots & \dots \\ 0 & \dots & \dots & 0 & r_n d_n \end{vmatrix}$$

where r_i are the eigenvalues of the matching matrix M, and the d_i are either zero or one.

Matching matrices are binary matrices, and this means that for any pair of generators i, j the (i,j) entry in the nth power of the matching matrix cannot be negative. This condition guarantees that, in the nth power J^n, the largest eigenvalue r_1 will completely dominate all other entries as n tends to infinity. r_1, the largest eigenvalue of the matching matrix, is known as the *spectral radius* of that matrix.

Now, consider very large patterns, where n is very large. Note that:

$$N_n = (e)(U)(J^n)(U^{-1})(e') = (a)(r_1)^n + \text{smaller terms}$$

where a is a positive constant. Thus, the following limit exists:

$$\lim_{n \to \infty} ((1/n)\log N_n) = \log r_1$$

One important characterization of pattern systems is the templeting index Q – a measure of the relative amount of templeting or of self-assembly in a particular pattern-assembly process – and is given by:

$$Q = S/S_{max}$$

From:

$$S = \log_2 N$$

and:

$$\lim_{n \to \infty} ((1/n)\log N_n) = \log r_1$$

$$S \sim n\log r_1$$

for large patterns; hence

$$Q \sim \log r_1/\log r_{max}$$

where r_{max} is the spectral radius of the least restrictive condensed matching matrix. The least restrictive matching matrix describes a set of topologically naive generators and is composed entirely of ones.

For a $g \times g$ symmetric matrix filled with ones the spectral radius is g and for large n:

$$Q \sim \log r_1 / \log g$$

As required, in completely templeted patterns $Q \longrightarrow 1$ because $r_1 = r_{max}$ when the matching matrix is completely nonrestrictive. And, in completely self-assembling patterns $Q \longrightarrow 0$ because $r_1 \longrightarrow 1$ when the matching matrix is highly restrictive. (For further details, see M.J. Katz & Y.-S. Chow (1985) 'Templeting and self-assembly' *J. Theoret. Biol.*, 113: 1–13.)

Parsimonious explanations

Although Nature is not necessarily parsimonious, scientific explanations should nonetheless be internally parsimonious. For example, suppose that we find the compound abc floating in the ocean. This compound is always made of the three molecules $\{abc\}$ strung together in the chain a-b-c. A configurational explanation for this compound would take the form:

$$MM \wedge T = AM$$

where the adjacency matrix is:

$$AM = \begin{array}{c} \\ a \\ b \\ c \end{array} \begin{array}{ccc} a & b & c \\ 0 & 1 & 0 \\ 1 & 0 & 1 \\ 0 & 1 & 0 \end{array}$$

The exact templet and matching matrices have not yet been chosen, but they must be 3×3 binary matrices in order to properly fit with the final adjacency matrix. The whole configurational explanation is parsimonious: it admits 'no more causes of natural things than such as are both true and sufficient to explain their appearances' (Newton).

In contrast, suppose that we have assumed that each molecule in the real world is actually an irreducible dimer and that the detailed molecular chain is:

$$(a_1\text{-}a_2)\text{-}(b_1\text{-}b_2)\text{-}(c_1\text{-}c_2)$$

Now, our adjacency matrix should be:

$$
\begin{array}{c}
 \quad a_1 \ a_2 \ b_1 \ b_2 \ c_1 \ c_2 \\
AM = \begin{array}{c} a_1 \\ a_2 \\ b_1 \\ b_2 \\ c_1 \\ c_2 \end{array}
\begin{array}{cccccc}
0 & 1 & 0 & 0 & 0 & 0 \\
1 & 0 & 1 & 0 & 0 & 0 \\
0 & 1 & 0 & 1 & 0 & 0 \\
0 & 0 & 1 & 0 & 1 & 0 \\
0 & 0 & 0 & 1 & 0 & 1 \\
0 & 0 & 0 & 0 & 1 & 0
\end{array}
\end{array}
$$

In some situations, this would represent a good working model of the real world and could thereby be incorporated into a parsimonious explanation. If, however, we have no evidence for this dimer model and especially if we have no way to test it, then the introduction of the dimer model merely adds superfluous elements. The model admits more causes of natural things than such as are both true and sufficient to explain their appearances, and the resulting explanation is not parsimonious.

In general, configurational explanations protect us against such nonparsimonious explanations because their adjacency matrices must come from scientific abstractions. The elements of a scientific abstraction have been set in a one-to-one relation with items of the real world, and in this way we are not allowed to introduce superfluous elements into our configurational explanations. Nonetheless, there remains the difficult problem of properly defining the items of the real world. In the example of the molecule abc, how do we know whether the dimer model is a good model or a superfluous model? How can we decide whether the basic item in the real world is the whole molecule a or whether the basic item is actually half of the molecule, either a_1 or a_2?

For this decision, the templet is critical. To see the role of the templet, consider first the matching matrix. A zero in the (i, j) slot in a matching matrix signifies that the two elements i and j are not connected, that they are distinct elements. If we can muster sufficient evidence to put zeros in the slots corresponding to the matching relations between two potential real world items, then those items can certainly be considered to be distinct. If there is some reason to believe that $m(a_1, a_2) = 0$, then we are justified in including both a_1 and a_2 as separate elements in a parsimonious configurational explanation.

Now, the fact that we are debating whether to consider a_1 and a_2 to be distinct items implies that the two elements are able to form an interconnection – i.e., that $m(a_1, a_2) = 1$ and that the matching matrix must have ones in the appropriate slots. Our dilemma arises because the two elements could be *permanently* connected: we wonder whether the elements are so integrally connected that they are, in fact, parts of the same operationally irreducible real world item. In this case, the dimer model is a superfluous complication that we are imposing on the real world, and explanations based on the dimer model are not parsimonious. It is here that the templet can help.

We have found that a_1 happens to be connected to a_2 in the particular pattern that we are examining. Thus, there must be a one in both the (a_1, a_2) and the (a_2, a_1) slots in our templet matrix. On the other hand, to say that a_1 and a_2 are distinct items means that there can be circumstances in which they are not connected and therefore that the (a_1, a_2) and the (a_2, a_1) slots in the templet matrix *could* contain zeros. If there is reason to believe that this is true and that a scientist could arguably create a templet with zeros in those slots, then both elements can be included in a parsimonious explanation. On the other hand, if the scientist cannot justify building a templet with zeros in the necessary slots, then a_1 and a_2 cannot be considered to represent distinct real world items and they cannot be included together in a parsimonious explanation.

How does a scientist justify his claim that he could put zeros in the (a_1, a_2) and the (a_2, a_1) slots of a templet? For this, rhetoric alone will not suffice. The decision is not one of 'consider the possibility that...'; instead, evidence must be presented. Herein lies the heart of parsimonious explanations: a parsimonious explanation considers that real world items are distinct only on the basis of actual evidence. For scientific explanations, parsimony means that an item is singular until there is real world data that is inconsistent with its singularity. Basically, then, a parsimonious explanation of the form:

$$MM \wedge T = AM$$

is one member of a family of configurational explanations with identical matching and adjacency matrices and for which, *in accord with real world evidence*, the zero templet (zeros in all slots) could occur.

**Maximal templets are required to distinguish between
a great many possible configurations**

An elemental principle of pattern construction is that to
create a complex pattern one must sometimes begin with a prepattern
of the same size; these patterns cannot be reduced to smaller
beginnings. In the terms of configurational explanations, if templets
are manifest in the form of patterns then certain pattern-assembly
situations require templets that are the same size as the final pattern
to be constructed. Here, 'size' means 'number of constituent elements'.

In other words, this principle contends that when a set of templets
can be used to distinguish any of N distinct objects – in this case,
the N distinct patterns permissible under a given matching matrix –
then the set must sometimes contain at least N different templets.
To demonstrate this principle, we first calculate the minimal number
of constituent elements necessary to construct a set of N distinct
templets, and to make this computation we must define the structure
of the generators. Our eventual goal is to directly compare the
number of elements in a pattern with the number of elements in its
templet – therefore, both the templet and the final pattern should
be built of comparable elements. Because we begin with a given
pattern, it is simplest to build the templet of the same units as our
pattern. Consider that the given pattern is composed of g types of
elements; P_n will be the number of patterns that can be constructed
from n or fewer of these elements. The number of patterns can be
ordered as:

$$P_1 \leqslant P_2 \leqslant P_3 \leqslant \dots$$

and in most cases the series can be strictly ordered:

$$P_1 < P_2 < P_3 < \dots$$

Suppose that we wish to construct some specific number N of
permissible patterns. N will have a place (i.e., P_n) in this ordered series,
where $P_{n-1} \leqslant N$. (We know that $P_n = N$, because an initial condition
is that n elements will form N distinct patterns.) When the series is
strictly ordered, $P_{n-1} < N$, and at least one of the N permissible
patterns must be composed of n or more elements. When the series is
strictly ordered – when the use of more elements always allows the
construction of more patterns – then P_{n-1} is always less than N; and,

except when a row or a column of our matching matrix is filled entirely with zeros, this will always be the case. Thus, as the principle contends, there are a great many patterns that require same-sized templets for their assembly.

Beyond the requirement for same-sized templets, the second principle of templeting reminds us that some patterns need *maximal* templets. Not only is a maximal templet the same size as its final pattern – a maximal templet specifies *all* of the connections in the final pattern, and it is the most explicit representation of the topology of that pattern. For an $n \times n$ adjacency matrix, there are $S_{max} = n^2$ independent slots, and the second principle of templeting claims that the assembly of certain patterns requires templets that can specify S_{max} matrix slots. Simple counting arguments lead ineluctably to this conclusion.

A templet must contain sufficient slots to distinguish between all of the patterns that can potentially be formed under any given matching matrix. Each templet slot can be used to distinguish two different patterns; hence, a templet with x independent slots can distinguish at most 2^x different patterns. If X is the number of patterns that can be distinguished by templets that have x or fewer independent slots, then:

$$X = \sum_{i=1}^{x} 2^i = 2^{x+1} - 2$$

When must a templet be as large as the final pattern that it is specifying? When must $x = S_{max}$? The number of different possible patterns that can be constructed under a given matching matrix is 2^S (where S is the minimal necessary templet size – the bits of information represented by the templet), hence:

$$X \leqslant 2^S$$

and:

$$2^{x+1} - 2 \leqslant 2^S$$

If:

$$x = S_{max}$$
$$= n^2$$

then the minimal necessary templet is as large as the final pattern when:

$$2^{nn} - 2 \leqslant 2^S$$

This equation is true whenever $S = n^2$ – i.e., whenever all of the independent slots in an $n \times n$ symmetric matrix are filled with ones. In other words, to distinguish between all of the distinct configurations that can be constructed from a completely nonrestrictive matching matrix, one must always use a templet that is as large as the final pattern. Maximal templets are always needed to construct patterns from topologically naive precursor elements. When 'universal laws' permit all possible patterns, the assembly of one particular pattern requires a maximal templet.

Nature has had available many paths as she built organisms, and the luxuriance of the possible blueprints is seen in the fact that natural templets are often *maximal* templets.

The ways chosen by Nature for making organic compounds are easy enough to rationalize – by hindsight. Biochemistry remains, by and large, an empirical science and predicting the course of even the simplest biochemical reactions, on whatever grounds, remains a venture of risk.

[K. Bloch (1976) on the evolution of a biosynthetic pathway. In: A. Kornberg, L.Cornudella, B.L. Horecker, and J. Oro (eds.). *Reflections on Biochemistry* Pergamon Press, Oxford, p. 145.]

Determinate (recurrent) maximal templets are the secret of life. They are the founts of the idiosyncratic complex patterns within organisms, they are randomness in captivity, and they are the reason that biology has been and will continue as a retrospective science. Biology is the true and original science of natural history.

A nonstatistical definition of randomness

Background
Traditionally, randomness has been defined in terms of probability: given an exhaustive set of events, if each event occurs with equal probability then any one occurrence can be called 'random'. A 'random sequence' of events is a string of these individual equiprobable events occurring independently. At the heart of Probability Theory, however, lies a dilemma – in his authoritative treatise on

probability, Feller [(1966) *An Introduction To Probability Theory And Its Applications*, 2nd edn. Wiley, NY, Vol. I, p. 29.] wrote:

The word 'random' is not well defined, but when applied to samples or selections it has a unique meaning. Whenever we speak of random samples of fixed size r, the adjective random is to imply that all possible samples have the same probability, namely, $1/(n^r)$ in sampling with replacement.

The major difficulty with such definitions is that they are based on the concept of equi-probable, and 'equi-probable' cannot easily be defined prospectively – that is, from *a priori* first principles. 'One of the most important problems in the philosophy of the natural sciences,' wrote Kolmogorov [(1956) *Foundations of the Theory of Probability*, 2nd English edn. (N. Morrison, transl.), Chelsea, NY, p. 9], 'is ... the essence of the concept of probability itself.'

To circumvent this difficulty, some mathematicians have called 'equi-probable' an elemental or primitive concept and have relied on our intuitive understanding of this notion [e.g., B.O. Koopman (1940) 'The bases of probability' *Bull. Am. Math. Soc.* **46**: 763–74; E. Borel (1965) *Elements of the Theory of Probability*, Prentice-Hall, Englewood Cliffs NJ, p. 16; B.V. Gnedenko (1968) 'Chapter 1. The concept of probability' *The Theory of Probability*, Chelsea, NY, pp. 21–87].

Other mathematicians have chosen to work only with mass or 'collective' phenomena [e.g., J.S. Mill (1846) *System of Logic, Ratiocinative and Inductive*, Harper Bros., NY, book III, chapter 18; R. von Mises (1957) *Probability, Statistics And Truth*, Macmillan, NY; R.A. Fisher (1966) *The Design of Experiments*, 8th edn. Hafner, NY; B.V. Gnedenko (1968) 'Chapter 1. The concept of probability' *The Theory of Probability*, Chelsea, NY, pp. 21–87.] and to define probabilities retrospectively in terms of frequencies. From the perspective of collective phenomena, one can say that two events had an equal probability of occurrence when, upon examining a large number of occurrences:

(a) The two events were found with an approximately equal frequency.

(b) The order of occurrences followed no simple pattern.

Both of these approaches leave a large gap at the foundation of any definition of 'randomness'. The first approach is not formal. A formal definition can be translated into standard symbols, whereas relying on purely intuitive notions means using ideas that cannot be written down formally. The second approach lacks penetration and insight; purely retrospective definitions can be formalized, but they are empirical and superficial. For a purely retrospective definition, we must accumulate a mass of experimental evidence, and we need propose no mechanisms describing either the inner workings or the interactions of the events themselves.

In contrast, ideal formal definitions are *deep* and thus they are *a priori*. Deep definitions model the inner structure of the events and of their interactions. They stem directly from intrinsic properties of the events themselves; therefore, they are *a priori* – that is, they can predict frequencies in advance.

In the case of 'randomness', we would also like our definition to go beyond *deep*; we would like the definition also to be determinate and not probabilistic. For 'randomness', we are attempting (in essence) to define 'equi-probable'; if our definition already depends on the concept of probability, then we will always be plagued with an internal circularity in our argument.

The construction of deep, *a priori* and determinate definitions of events is equivalent to the precise description of machinery that will generate these events. A well-defined machine is determinate, the detailed description of a machine can provide an *a priori* prediction of its operation, and by outlining the constituent parts of the machine the definition becomes deep.

For completely determinate events – that is, for events that occur completely predictably – the act of definition poses few philosophical problems. Determinate events can be directly modeled by ideal machines, because ideal machines are by their very natures determinate. Here, we need only describe the underlying mechanisms of the appropriate machine as well as we are able. 'Equi-probable' and related concepts cast no shadows in this realm – operations defined in terms of determinate machines can, in theory, be defined determinately, without resorting to probabilities.

On the other hand, for stochastic events – that is, for events generated by processes that cannot be completely predicted in advance – a precise definition of the underlying machinery always includes some sort of probabilistic statement. At some point, we find ourselves saying: 'This apparatus will produce one of a number of events, and the probability of event (a) is ...' In this vein, for definitions of random events, the classic description will include a statement like: 'This apparatus produces one of a number of events, and the production of each event is equi-probable.' For instance, R.A. Fisher [(1966) *The Design of Experiments*, 8th edn. Hafner, NY, p. 51] wrote that randomization is a 'physical experimental process ... [that] ensures that each variety has an equal chance of [occurrence].' With stochastic events, we run into a philosophical problem: randomness is intimately tied to stochasticism. Is it possible to define randomness without probability?

Randomness and complexity

The long and thoughtful history of struggles with this problem suggests that the concepts 'probability' and 'randomness' inescapably include both collective events and unpredictable events. Nonetheless, we would like to build definitions that are independent of probability and definitions in which the role of 'unpredictability' is as circumscribed and as well defined as possible. In the mid 1960s, a major step toward this end was taken by A.N. Kolmogorov [(1965) 'Three approaches to the quantitative definition of information' *Problems Inform. Transmission* 1: 1–17; (1968) 'Logical basis for information theory and probability theory' *IEEE Trans. Inform. Theory IT-14*: 662–4] and G.J. Chaitin [(1966) 'On the length of programs for computing finite binary sequences' *J. Assoc. Computing Mach.* 13: 547–69; (1977) 'Algorithmic information theory' *IBM J. Res. Develop.* 21:350–9]. These mathematicians pulled 'randomness' away from 'probability' by forging a fundamental connection between 'randomness' and 'complexity', which is a determinate concept. Specifically, Kolmogorov and Chaitin showed that certain patterns (binary strings) that were maximally complex and that were determinately generated also met all of the statistical criteria of randomness.

Maximally complex binary strings are very heterogeneous patterns of ones and zeros. In contrast to 'random', however, 'maximal complexity' can be defined completely determinately, in terms of precisely formulated machines. Thus, although they are statistically indistinguishable from random, maximally complex patterns need not meet all of the key criteria for truly random strings. In particular, the maximally complex patterns examined by Kolmogorov and Chaitin are entirely predictable: their maximally complex strings are those strings that are always output by a determinate machine (a computer) when the machine is given particular (albeit, complex) well-defined input strings.

In this way, Kolmogorov and Chaitin separated two independent components of randomness. On the one hand, there is maximal complexity – random patterns are highly heterogeneous, and they are thereby very complex. On the other hand, there is unpredictability – random patterns are generated by stochastic processes, and, unlike the maximally complex strings put out by determinate computers, truly random patterns are not absolutely predictable.

Templeting definition of randomness

A. Complete templets and complexity

Random patterns have two essential features: they are maximally complex, and they are generated stochastically. How are these two features expressed in a configurational explanation?

First, let me examine complexity. A complex pattern is a pattern that requires a substantial templet for its fabrication, and a maximally complex pattern requires a maximal or complete templet. For complex patterns, the 'universal assembly laws' are insufficient to uniquely determine one particular pattern configuration; some configurational information must also be introduced in the form of a prepattern. For maximally complex patterns, the prepattern is a complete templet – that is, it specifies the entire pattern. As Kolmogorov and Chaitin demonstrated, maximally complex binary string patterns can only be fabricated from precursor strings of the same size, and this result can be extended to the general principle that all maximally complex patterns require templets of the same size – these are complete templets, templets necessarily as large as the patterns themselves.

Random patterns are fabricated from high entropy systems; they require complete templets for their construction. Nonetheless, this need for complete templets is not in itself sufficient to define random patterns: 'randomness' demands an additional property. For example, some natural patterns, such as many DNA sequences, require complete templets and are maximally complex, but they are, nonetheless, not random. Truly random patterns must also have another characteristic, the property that we call 'unpredictability'. This means that random patterns must also be generated stochastically.

B. Ephemeral templets and stochasticism

Like *complexity*, *unpredictability* – the final distinguishing feature of random patterns – is also determined by the patterns' templets. All maximally complex patterns have maximal templets, templets that completely specify their configurations. For random patterns, however, the templets have an additional characteristic: they are ephemeral. Templets for truly random patterns have no long-term coherence, and they must be recreated each time a pattern is fabricated. A stochastically-generated pattern is one for which the templets have no permanence; each 'run' of a stochastic process generates a templet anew.

De novo templets characterize stochastic pattern-assembly, and maximal *de novo* templets characterize random pattern generators. Wherein, then, lies the 'unpredictability' of randomness? 'Unpredictability' is a second feature of stochastic pattern generating processes – besides creating the templets anew with each run of the machine, stochastic processes have a special relation to the observer. This relation can be called 'unpredictability'.

'Unpredictability' comes when we, the observers do not have sufficient *a priori* knowledge to definitely describe the particular templet that will be created for a given 'run' of the pattern-generating process. It may be, as in the case of flipping a coin, that it is impractical for us to make a definite prediction because too many events and too many influences conspire together in the creation of the templet. It may be, as in the case of the behavior of a child, that the templet is built with characteristically human capriciousness. Or, it may be, as in the case of subatomic particles, that the underlying machinery is forever beyond the resolution of human science. In any case, 'unpredictability' is always synonymous with our lack of complete foreknowledge of the operant templet.

This lack of foreknowledge can be described more precisely in terms of configurational explanations. Here, we need to formally describe ephemeral templets:

A templet can itself be considered to be a pattern; thus, it too can be analyzed as a configurational explanation. A determinate (a permanent) templet is a pattern for which the same configurational explanation can always be constructed. At each usage, identical precursors – the same matching matrix M' and the same templet matrix T' – are always invoked to fabricate the operant templet T. If T is the operant templet matrix, and if M' is the matching matrix for T and T' is the templet matrix for T, then the configurational explanation for a determinate templet is simply:

$$M' \wedge T' = T$$

In contrast, a stochastic (an ephemeral) templet is a pattern that is effectively recreated at each usage and for which different precursors may be employed in the various recreations. Although one of the precursors – the set of elements (as represented by the matching matrix M') – is always the same, the other precursor – the prepattern (the *templet's* templet matrix T') – can vary.

For stochastic pattern-assembly systems, the process that constructs the templet T has at its disposal a number of different prepatterns (templet matrices: $T'_1, T'_2, \ldots, T'_n$), and we do not possess sufficient advance information to predict which prepattern will actually be used at any one time. If T is the operant templet matrix, and if M' is the matching matrix for the templet T and T' is the templet matrix for the templet T, then a number of different configurational explanations are available for a stochastic templet:

$$M' \wedge T'_1 = T$$
$$M' \wedge T'_2 = T$$
$$M' \wedge T'_3 = T$$
$$\ldots\ldots\ldots$$
$$M' \wedge T'_n = T$$

A nonstatistical definition of randomness
In summary, a random pattern has two key characteristics:

(a) It is maximally complex – that is, it is fabricated using a complete templet.

(b) It is unpredictable – that is, the complete templet is recreated *de novo* at each usage, and we have insufficient advance knowledge to define the actual templet appearing at any one particular time.

This definition follows in the footsteps of George Boole who, in the mid nineteenth century, phrased probability theory in the rigorous and general language of his Calculus of Logic and who wrote in 1854:

Probability is expectation founded upon partial knowledge. A perfect acquaintance with *all* the circumstances affecting the occurrence of an event would change expectation [stochasticism] into certainty [determinism], and leave neither room nor demand for a theory of probabilities.

[G. Boole (1940) *Collected Logical Works. Vol. II. The Laws of Thought (1854)*, Open Court Publ., Chicago, p. 258.]

Example: coin flipping, a random pattern generator

As an illustration, consider coin flipping, the archetypic random pattern generating system. How can coin flipping be formalized as a configurational explanation? In the simplest situation, you toss a coin into the air and allow it to fall onto a table. There are two possible outcomes – Heads or Tails – and each of these final patterns can be abstracted as a simple graph. 'Tails' will be symbolized by '$H - t$', meaning that the Heads (H) side of the coin is adjacent ($-$) to the table (t); whereas, 'Heads' will be symbolized by '$H\ t$', meaning that the Heads side of the coin is *not* adjacent to the table.

A configurational explanation is written as the concatenation of three matrices. As an adjacency matrix, A, the pattern $(H - t)$ representing 'Tails' is:

$$A = \begin{array}{c} \\ H \\ t \end{array} \begin{array}{cc} H & t \\ 0 & 1 \\ 1 & 0 \end{array}$$

and the pattern $(H\ t)$ representing 'Heads' is:

$$A = \begin{array}{c} \\ H \\ t \end{array} \begin{array}{cc} H & t \\ 0 & 0 \\ 0 & 0 \end{array}$$

The matching matrix M, for either of these patterns, is completely nonrestrictive: all possible interconnections can potentially occur between H and t. The matching matrix is:

$$M = \begin{array}{c} \\ H \\ t \end{array} \begin{array}{cc} H & t \\ 0 & 1 \\ 1 & 0 \end{array}$$

124   *Appendix: A pattern theoretic formalization*

Finally, to build a complete configurational explanation, $M \wedge T = A$, one of two possible templets must be specified. Either:

$$T = \begin{array}{c|cc} & H & t \\ \hline H & 0 & 0 \\ t & 0 & 0 \end{array}$$

because:

$$\begin{array}{cc} 0 & 1 \\ 1 & 0 \end{array} \wedge \begin{array}{cc} 0 & 0 \\ 0 & 0 \end{array} = \begin{array}{cc} 0 & 0 \\ 0 & 0 \end{array}$$

for Heads ($H\,t$). Or:

$$T = \begin{array}{c|cc} & H & t \\ \hline H & 0 & 1 \\ t & 1 & 0 \end{array}$$

because:

$$\begin{array}{cc} 0 & 1 \\ 1 & 0 \end{array} \wedge \begin{array}{cc} 0 & 1 \\ 1 & 0 \end{array} = \begin{array}{cc} 0 & 1 \\ 1 & 0 \end{array}$$

for Tails ($H - t$).

The complete configurational explanation for coin flipping demonstrates that the final pattern (either $H\,t$ or $H - t$) can appropriately be called 'random' because it fulfills the two essential criteria for randomness:

(a) It is maximally complex; in each case, a complete templet must be invoked – that is, the necessary templet specifies the entire pattern.

(b) It is unpredictable; in each case, the pattern is generated by a stochastic process – namely, a process that recreates the templet anew in each run and a process about which we have incomplete foreknowledge. Here, the operant templet is always a *de novo* concatenation of all the forces that go into flipping a coin, and it is impractical for us to assess (especially, in advance) the sum of those forces for any particular flip.

Coin flipping produces patterns that are maximally complex and that are unpredictable; and it is precisely in these two ways that coin flipping produces random patterns. Randomness stems from ephemeral unpredictable templets.

Quotation index

Subject index

adjacency matrix (final pattern) 100–1
art is templeted 95–7
automata, finite 40–6, 102–8

brain 11–12, 54

coin flipping 123–4
complexity
  as difficulty of fabrication 74
  as heterogeneity of configuration 73–4
  as heterogeneity of content 73
  as physical size 73
  history of definition 75–7
  of formulas 76–7, 87–90
continuum math versus discrete math 8–9

determinism 20–2
discrete phenomena 7–9
DNA 56–7

elegance 12–14
entropy 91–5
evolution 63–6
excesses in nature 51–5
explanatory unit 86–7
extremum principles 51–5

'how' questions 18–19

irreducible complexity 26–7

life
  definition 81–4
  origin 65–6

machines 39–40
mappings, one-to-one 6–7
matching matrix ('universal assembly laws') 99–100

nonanalytic expression 87–90

Occam's razor 49–55
one-to-one mappings 6–7
ontogeny (development) 62–3
order, not created *ex nihilo* 3, 47–8
'order' as a deceptive word 93–5
origin of life 65–6

parsimony
  in explanations 49–51, 111–13
  in nature 49–55
pattern theory 90–101
phylogeny (evolution) 63–6
polarized processes 91–5
predictions 22–4

quantum theory, lack of depth 16, 21–2

random patterns 1, 78–81, 116–24

scientific explanations
  algorithm 27, 68–71
  as automata 40–6, 102–8
  information content 38
self-assembly, examples 60–3
simple rules, theories, and explanations 50–1, 66–7, 90–1
stick figure abstraction 5–6, 9–10, 24–7
stochastic processes 78–81, 93–5, 121–2

templeting, two principles 48–58
templets
  ephemeral 121–2
  equivalent 36
  etymology 34–5
  examples 57–8
  maximal 37, 114–16
  minimal necessary size 37–9, 114–16
  unique 35–6

Subject index   127